COAL GASIFICATION

Existing Processes and New Developments

COAL GASIFICATION

Existing Processes and New Developments

H-D Schilling, B Bonn, U Krauss

Introduction by W Peters

Published by Graham & Trotman

Translation from German by Cave Translations Ltd, London, arranged by the Commission of the European Communities, Directorate-General Information Market and Innovation

Published in English in 1981 by Graham & Trotman Limited, Bond Street House, 14 Clifford Street, London W1X 1RD

ISBN 0 86010 263 7

Originally published in German as Rohstoffwirtschaft International Band 4. Kohlenvergasung: Eine Basisstudie über bestehende Verfahren und neue Entwicklungen

ISBN 3-7739-0240-9

Printed in England by Spottiswoode Ballantyne Ltd.
Colchester and London

Foreword to the Second edition

The first edition of this book was rapidly exhausted. The continued demand, together with developments since 1976, have prompted us to publish a second and thoroughly revised edition. The form of presentation used successfully in the first edition has been retained, but all the data have been checked and brought up to date where necessary. Some further gasification processes have been added to the list, and examples of underground gasification are now dealt with in a separate section. Conversely, in some areas processes which appear to be of limited application or have been abandoned are no longer described in detail, in order to save space. Revolving grate generators and slagging gasifiers, which have already been thoroughly reviewed in the older literature and have recently been attracting renewed interest, are grouped together in separate chapters.

In response to a number of requests, the Introduction has been enlarged to include a generally comprehensible account of the complex interplay in coal gasification between raw material properties, chemical reactions, processing parameters and conditions of industrial application. A table summarizing all the important characteristic data of gasification processes has also been added to the Introduction, in order to provide the reader with a rapid survey.

Our thanks are due to those readers of the first edition of this book whose comments have been helpful in revision of the text for the second edition. Especial thanks go to Dr H.W. von Gratkowski, who read through the manuscript of the second edition and made numerous suggestions for its improvement.

Professor Dr Werner Peters
Managing Director
Steinkohlenbergbauverein
Bergbau-Forschung GmbH

Note on Units

SI units are used preferentially.
All volumes indicated by m^3(iN) are at standard conditions, viz. a temperature of 0°C (273 K) and pressure of 1.013 bar.
H_o is the higher heating value (HHV).

Essen, February 1979

Contents

1. Introduction 5

2. Fundamental reaction chemistry of coal gasification 10

3. The various routes to coal gasification 22

4. Main lines of development in new process design 33

5. Explanation of the form of presentation 39

6. Survey of gasification processes 46

A. Processes currently in commercial operation 51

1. Revolving-grate generator processes 52
2. Lurgi pressure gasification process 71
3. Winkler process 85
4. Koppers-Totzek process 96

B. Processes not currently in commercial operation and processes under development 107

1. Hot-blast gas processes 108
2. Slagging-bed generator processes 120
3. British Gas Corporation high-pressure slagging gasifier process 129
4. Hygas process 138
5. Cogas process (COED) 150
6. CO_2 acceptor process 160
7. Bi-Gas process 171
8. Ash-agglomerating process 179
9. Synthane process 185
10. U-gas process 193
11. Texaco process 200
12. Rummel-Otto process 210
13. Ruhrgas vortex chamber process 220
14. Combustion Engineering process 227
15. Westinghouse process 233

C. Gasification processes still at the planning stage (small-scale experiments) 241

1. Exxon process 242
2. Hydrane process 248
3. Molten salts process (Kellogg) 254
4. Coal gasification in molten iron (Atgas process) 259

D. Gasification of coal using nuclear process heat 267

1. Steam gasification of lignite 270
2. Steam gasification of hard coal 279
3. Hydrogasification of lignite 289

Indirect coupling of nuclear heat via steam reforming of methane in a tube furnace 290

Coupling of nuclear heat via pre-heating of the gasifying medium 302

4. Hydrogasification of hard coal 310

E. Underground gasification 311

1. Introduction

Gaseous fuels are among the most sought-after raw materials and energy sources in the energy economy. Their excellent properties as regards environmental protection and ready transportability, combined with their wide range of uses as energy sources and chemical feedstocks, mean that their importance in both sections of the market is likely to increase rather than decrease in the long term.

However, in the face of the state of world reserves of primary energy sources and the steadily rising demand for energy, it has become necessary to replace natural gas and oil by other energy sources to an increasing extent. It is estimated that a substitution of this kind will need to take effect before the turn of the century. The conversion of coal into a gaseous fuel and raw material will be of decisive importance in attempts to secure the future of energy supplies.

Although the gasification of coal was practised as long ago as the 19th century and was developed to an advanced stage of technology in the first half of the present century, especially in Europe, nowadays it is of importance only in a few countries with special economic conditions. In the Federal Republic of Germany in 1975, although 20.4% of total gas production came from coal only 0.3% of this amount was obtained by gasification. (In 1957 by comparison, 92.78% of total gas production came from coal, of which 13.7% was obtained by gasification.) Let us therefore briefly examine the past and the probable future significance of gasification in the production of gas from coal.

Three main routes can be followed for producing gas from coal on a large industrial scale. Processes involving pyrolysis are the most important, both quantitatively and economically. When coal is pyrolysed its volatile constituents are cracked to give gaseous and liquid products, which can be separated and worked up to provide marketable by-products. These processes represent a de-gassing of the coal, in which the greater part of the feed material remains as solid coke.

Large quantities of gas are also produced in blast furnaces. The reaction of the ore with coke in the presence of air produces considerable quantities of carbon monoxide, which leave the furnace as stack gas. This gas is combustible and is used as heating gas. Because of its low calorific value it is not, however, economic to transport the gas over long distances.

In both of the above instances the gas is produced merely as a by-product from other processes, so that the quantity of gas available depends primarily on the demand for the main product. In contrast to this the third route, which is the complete conversion of coal into gas, is aimed exclusively at the production of gas, and it is this route alone that constitutes the subject of this book.

In earlier decades coal gasification served not only to release gas supplies from dependence on coke production, but also to make pure synthesis gas available to the chemical industry. Thus in 1957 in the Federal Republic of Germany (excluding the Saar region) gas manufactured by coal gasification amounted to 12.8% of total gas production, or 113.4×10^{15} J/a. Of this, about 75% came from industrial generators which produced both synthesis gas and fuel gases, and about 25% from coke ovens and public gas works.

With the rapidly increasing supplies and steadily falling prices of petrochemical feedstocks in the 1950s and 1960s, the chemical industry and most district gas works changed over to the gasification of oil and naphtha, which made gas production cheaper and technically simpler. As a result the importance of coal gasification sank sharply, and in the Federal Republic of Germany the production of producer gas fell continuously up to 1975 to about 0.06% of total gas production (1.48×10^{15} J/a in absolute terms). Of this amount only one sixth was used for public gas supplies; the remainder was used in industry or for bottom firing in gas works and coke ovens. The proportions of coke-oven gas and stack gas decreased to a much smaller extent, for the reasons mentioned earlier: total production of gas from coal fell by only about 35% between 1957 and 1975.

The role of coal gasification in filling the gap between the demand for gas and the production of coke-oven gas, and also its importance for the chemical industry, lost almost all significance with the introduction of oil and naphtha cracking and with the expansion in the use of natural gas, which has been available for some years in large quantities in the Federal Republic of Germany at low cost. Total gas consumption in the Federal Republic of Germany has more than trebled during the last twenty years, but this has been strongly biassed in favour of natural gas. In the second half of the 1950s natural gas was still relatively unimportant, accounting for some 2% of total gas consumption, but by 1975 it played a leading part, taking 70.1% of the market.

As things stand at present, there are no obstacles blocking this rapid advance of natural gas. In contrast to coke-oven gas, natural gas can be transported economically even over long distances, and the national distribution system can therefore still be considerably expanded. However, the growing trend towards natural gas presents two problems. In the first place, reserves of natural gas are limited: proven domestic resources of natural gas amount to only rather more than 12 times present domestic consumption, and home production (1975: 45% of total consumption of natural gas) is already declining. As a result Germany's dependence on energy imports will increase further, and in view of the foreseeable exhaustion of world gas reserves this may lead to serious economic difficulties.

In the second place, a further increase in the consumption of natural gas will encompass, in particular, an increase in the proportion used for heating. This is accentuated by the fact that in the Federal Republic of Germany the construction of additional power stations fired by natural gas is being authorized on grounds of energy policy. Although the increased use of gas in place of other fuels in private households and by small consumers is in itself desirable, because it is environmentally acceptable and because the fact that it is consumed in easily controlled and monitored appliances encourages economy of use, this trend does however, aggravate seasonal fluctuations. The extremely high cost of the distribution network means that it must be operated for as much of the time as possible on base load, but this can no longer be achieved economically by gas storage if the fluctuations are very great. Furthermore, the required geological conditions for the storage of large quantities of gas do not always exist near the main centres of large consumption.

This problem is only too evident in many parts of the USA, where very real shortages have already become apparent. More than a decade ago, therefore, the major companies instituted research and development programmes aimed at producing a substitute natural gas (SNG). These investigations were at first directed towards the production of high BTU gases from oil. Following the steep rise in the price of crude oil and in line with "Project Independence", much greater emphasis has now been placed on coal as a starting material.

Since the events of 1973 the further development of coal gasification is also being pursued in Europe, especially in the Federal Republic of Germany, with substantial financial support from public funds. The main aim has been to develop processes for making synthesis gas, reducing gas for direct reduction, and cheap, clean fuel gas for use by industry and in power stations. In this way it should be possible to retain the technical advantages associated with the use of gas, without consuming already scarce natural gas and oil or secondary products derived from them.

In view of the new problems and potential applications associated with gas-from-coal technology, it seems opportune to give a comprehensive account of the technical requirements and development status of coal gasification. The purpose of this book is to provide a description of gasification processes, backed up as far as possible by comparable numerical data, which will aid objective discussion of the advantages and disadvantages of coal gasification for particular applications, at least as regards the technical aspects.

References: Annual Review of the Development of the Gas Industry in the Federal Republic of Germany, published by the Bundesministerium für Wirtschaft (Federal Ministry of Economic Affairs), Report III C 4; also published in the journal Gas- und Wasserfach (Gas).

2. Fundamental reaction chemistry of coal gasification

The gasification of coal can be summarized in five simple basic chemical reactions:

(a) Gasification with oxygen or air (partial combustion)

$$C + 1/2\ O_2 = CO \qquad \Delta H = -123.1\ kJ/mol$$

(b) Combustion with oxygen

$$C + O_2 = CO_2 \qquad \Delta H = -405.9\ kJ/mol$$

(c) Gasification with carbon dioxide (Boudouard reaction)

$$C + CO_2 = 2\ CO \qquad \Delta H = 159.7\ kJ/mol$$

(d) Gasification with steam (water gas reaction)

$$C + H_2O = CO + H_2 \qquad \Delta H = 118.9\ kJ/mol$$

(e) Gasification with hydrogen (hydrogasification reaction)

$$C + 2\ H_2 = CH_4 \qquad \Delta H = -87.4\ kJ/mol$$

The following reaction is of importance for making gas mixtures with a specified CO/H_2 ratio for use in subsequent syntheses:

(f) Water-gas shift reaction

$$CO + H_2O = H_2 + CO_2 \qquad \Delta H = -40.9\ kJ/mol$$

An increase in calorific value can be produced by methanation:

(g) Methanation

$$CO + 3\ H_2 = CH_4 + H_2O \qquad \Delta H = -206.3\ kJ/mol$$

The enthalpies of reaction have been given for standard conditions (298 K; 1.013 bar). The standard enthalpy of formation of carbon has been taken as $\Delta H^o = 12.5$ kJ/mol. This allows for the fact that the carbon in coal differs from graphitic carbon.

The equations (a) to (g) give the stoichiometry and energy change of the reactions which take place in gasification, but provide no information on the completeness of the reaction or the way in which it proceeds. This cannot be obtained until thermodynamic and kinetic laws and the conditions for stationary operation are also incorporated.

Like all chemical reactions, those given in equations (a) to (g) tend to a state of equilibrium, which may be one-sided. This means that after a sufficient reaction time, depending on the temperature and pressure, a fixed relation exists between the amounts of starting materials and end products. This relation is expressed by the equilibrium constant K_p. The subscript p indicates that the quantities of the individual components are expressed in terms of their partial pressures. The thermodynamic relations

$$\Delta G^o(T) = -RT \ln K_p(T)$$

and $$\Delta G^o(T) = \Delta H^o(T) - T\,\Delta S^o(T)$$

(ΔG^o = standard free enthalpy of reaction; ΔH^o = standard enthalpy of reaction; ΔS^o = standard entropy of reaction; standard state: T; 1.013 bar)

can be combined with empirical expressions for calculating the temperature-dependent specific heats to give expressions of the following form for K_p:

$$\lg K_p(T) = \frac{a}{T} + bT + cT^2 + d \lg T + e$$

The constants a to e are interpolated values derived from measurements, and can be taken from published tables (1). The following list gives the expressions for K_p corresponding to the individual reactions (c) to (g), together with the stoichiometric relations needed to calculate the proportions of the component gases at equilibrium:

Reaction c:

$$\frac{K_p}{P} = \frac{x^2_{CO}}{x_{CO_2}} \qquad (x_{CO} + x_{CO_2} = 1)$$

Reaction d:

$$\frac{K_p}{P} = \frac{x_{CO}\ x_{H_2}}{x_{H_2O}} \qquad (x_{H_2} + x_{CO} + x_{H_2O} = 1)\ ;\ x_{H_2} = x_{CO}$$

Reaction e:

$$P \cdot K_p = \frac{x_{CH_4}}{x^2_{H_2}} \qquad (x_{CH_4} + x_{H_2} = 1)$$

Reaction f:

$$K_p = \frac{x_{CO_2} \cdot x_{H_2}}{x_{CO} \cdot x_{H_2O}} \qquad (x_{CO_2} + x_{H_2} + x_{CO} + x_{H_2O} = 1)$$

Reaction g:

$$P^2 \cdot K_p = \frac{x_{CH_4} \cdot x_{H_2O}}{x_{CO} \cdot x^3_{H_2}} \qquad (x_{CH_4} + x_{H_2O} + x_{CO} + x_{H_2} = 1)$$

In these expressions P is the total gas pressure and x the mole fraction of the gas indicated. These quantities are related to the partial pressure P_A by the expression $x_A = P_A/P$. The volume percentage of each gas is obtained by multiplying the appropriate mole fraction by 100. The data can be used to calculate the dependence of the equilibrium composition on temperature. At high gas pressures real gas imperfections must be taken into account by introducing fugacities. The resulting equilibrium curves are shown in Figures 1.1 to 1.5. In each case the ordinate represents the proportion by volume of the final product.

Figures 1.1 to 1.5 show that in the most important gasification reactions changes of pressure or temperature within relatively narrow limits can produce significant shifts of the equilibrium composition towards either the starting materials or the desired end products.

<u>Figures 1.1-1.5 Equilibrium concentrations of the reaction products of reactions (c) to (g) as a function of the reaction temperature</u>

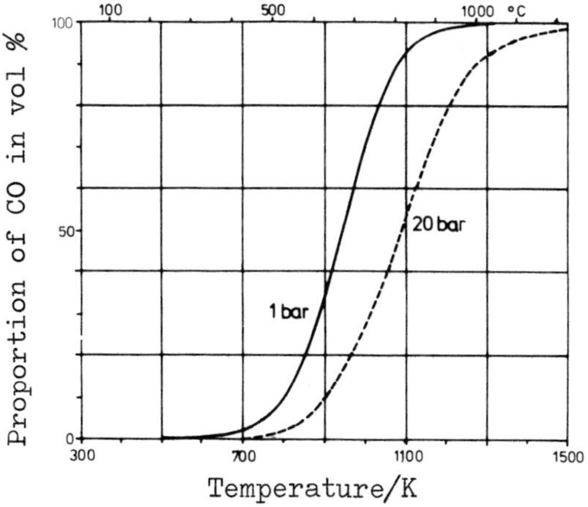

<u>Figure 1.1</u> CO/CO_2 equilibrium (reaction c)

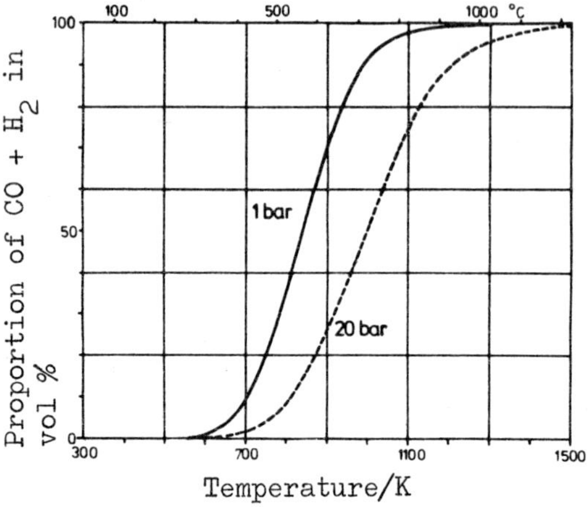

Figure 1.2 $H_2O/CO/H_2$ equilibrium (reaction d)

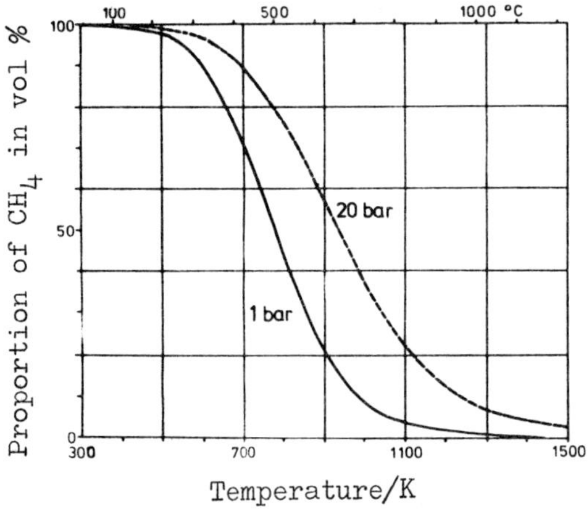

Figure 1.3 CH_4/H_2 equilibrium (reaction e)

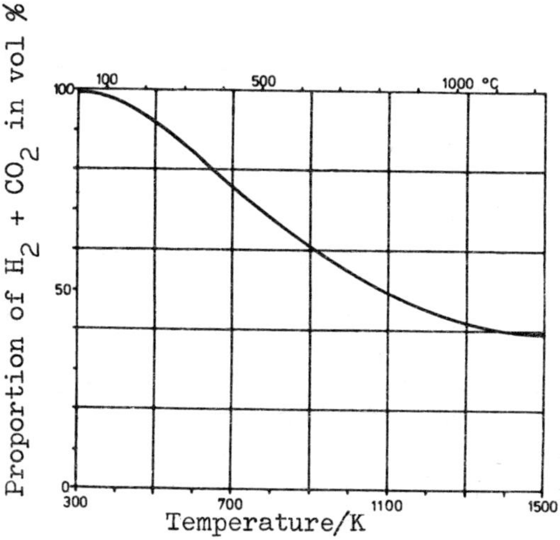

<u>Figure 1.4</u> $CO/H_2O/CO_2/H_2$ equilibrium (reaction f)

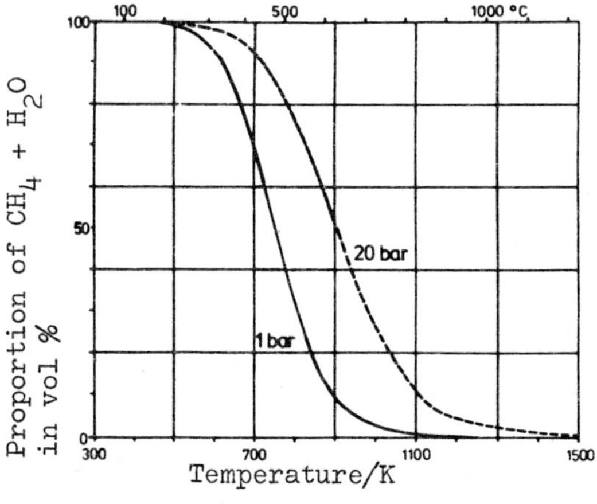

<u>Figure 1.5</u> $CO/H_2/CH_4/H_2O$ equilibrium (reaction g)

The shape of the curves shows that we can distinguish between two groups of reactions. Whereas for reactions (c) and (d) an increase of temperature favours the formation of the end products, for reactions (e), (f) and (g) raising the temperature displaces the equilibrium composition towards the starting materials. A knowledge of equilibrium curves is thus very important in relation to the product required and the choice of reaction temperature.

The reactions in question can be further subdivided if we consider the effect of pressure on the equilibria. Reaction (f) is almost pressure independent: the same equilibrium curve is always obtained if we disregard the departure of the individual components from ideal behaviour. All the other equilibria, on the other hand, are affected by pressure. Here we can distinguish between reactions (c) and (d), for which an increase of pressure displaces the equilibrium towards the starting materials, and reactions (e) and (g), in which raising the pressure favours the formation of the end products.

To illustrate the magnitude of the effect of pressure, Figures 1.1 to 1.5 each also contain the equilibrium curve for a higher pressure (20 bar; cf also (2)).

In the light of these relations it is clear that the formation of methane in the hydrogasification of coal is favoured by a high hydrogen pressure and a relatively low reaction temperature. On the other hand, the formation of carbon monoxide by gasification will occur to a large extent only at high temperatures and normal pressure, since the equilibrium then lies almost completely on the side of carbon monoxide.

In designing a gasification process it is necessary to take into account the equilibrium position of the reactions concerned in the range of pressures and temperatures chosen. It is possible in principle to combine all the reaction equations (c) to (g) and also the corresponding equilibrium expressions so as to calculate an equilibrium composition that allows for all the individual reactions. However, such an approach is invalid since it assumes that all the equilibrium relations can be applied simultaneously and on an equal footing. This assumption does not correspond to reality. In a gas generator there exists a spatial distribution of temperature, pressure and concentration. The equilibrium composition assumes values that vary spatially with pressure and temperature. Not all the reactions occur simultaneously; some are consecutive to others. This is illustrated by Figure 2, which shows the measured gas composition at different levels in a self-charging reactor in which coal is gasified with moist air (3).

The atmospheric oxygen is first converted to carbon dioxide in a shallow combustion zone. The decomposition of steam and the formation of carbon monoxide occur to an appreciable extent only when the oxygen has been used up and the coal and gas have become hot.

The actual composition of the gas is critically dependent on the rate at which equilibrium is attained, as described by the reaction kinetics. This chemical reaction velocity, which can only be measured experimentally, has not been taken into account so far in our deliberations on the course of the reactions, but is implicitly contained in the curves in Figure 2. For all gasification reactions it is very low at low temperatures, and increases exponentially with temperature. A distinction must therefore be made between the gas composition corresponding to thermodynamic equilibrium and the composition which can be reached during a given residence time taking kinetics into account. The equilibrium values for the proportions of products in a gas mixture represent only limiting values and correspond to the limiting case of an infinitely long reaction time under constant conditions. Each of the

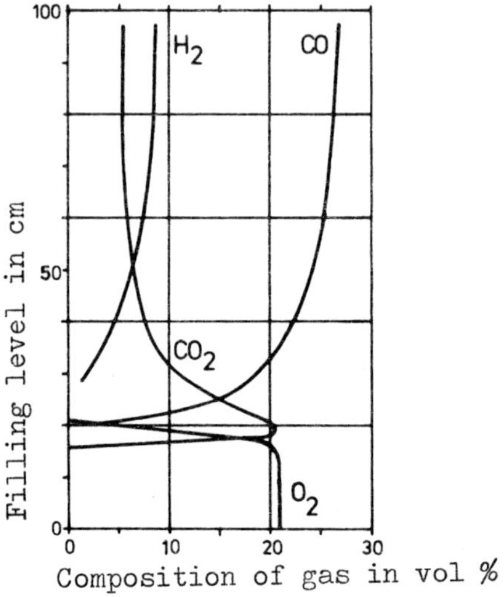

<u>Figure 2</u> Formation of producer gas in a shaft charged with coal (height of ash layer: 16 cm) (3)

reactions (a) to (g) shows individual kinetic behaviour, and therefore takes place at a different rate under the same reaction conditions. This precludes any assumption that all the reactions will proceed substantially to equilibrium simultaneously. In addition, it is necessary to take into account the spatial distribution of concentrations, temperature and pressure mentioned above.

In industrial gasification processes the flow rate of the reacting gas mixture and the reaction volume fix a residence time. There is thus a definite time available for the reaction to proceed. The reaction velocity is determined by the choice of pressure and temperature. These boundary conditions define a stationary gas composition (cf also (4)). The task of chemical plant design is to construct a gas generator on the basis of these boundary conditions which will produce a gas mixture approximating as closely as possible to the desired composition.

For the reasons given above the proportions of the individual components in the gas produced do not correspond to the requirements of thermodynamic equilibrium. Owing to the effect of thermodynamic and kinetic factors and the stationary state conditions, with the added complication of the spatial distribution of concentrations, temperature and pressure, it is not possible to predict beforehand the exact composition of the gas produced. Approximate values can, however, be calculated.

Some additional factors have not yet been taken into account. In the first place some of the reaction products may have an inhibiting effect on the velocities of other reactions. Secondly, in an industrial process using coal or coke allowance has to be made for the effect of the ash on the reaction velocity.

Under certain circumstances the ash may catalyse some of the gasification reactions selectively. This means that at a given temperature and pressure the velocities of some reactions may exceed the velocities they attain in the absence of catalysts under otherwise identical conditions. Catalysis has no effect on thermodynamic relations, including the position of equilibrium, but it can influence the stationary concentrations.

Since the ash content and composition vary from one coal seam to another, and even within the same seam, this introduces further difficulties into the accurate prediction of the actual gas composition, since different ash compositions will give rise to gases of different composition. Nevertheless, when particular ranks of coal are used in industrial reactors it is possible to give average values for the composition of the raw gas. For the reasons given above these values will fluctuate within certain limits, but can serve as a useful guide.

References

(1) Tabulated values for the constants a, b, c, etc

a. Landolt-Börnstein, Zahlenwerte und Tabellen (Numerical values and tables), 6th edition, Part 4, Berlin/Göttingen/Heidelberg, 1961

b. Nat. Bureau of Standards, Circular 500 Selected values of chemical thermodynamic properties, Washington, 1952

c. C.E. Wicks and F.E. Block Thermodynamic properties of 65 elements - their oxides, halides, carbides and nitrides Bulletin 605, Bureau of Mines, 1963

d. R.C. Weast (Ed) Handbook of Chemistry and Physics Chemical Rubber Co.

(2) S. Peter, G. Woyke and G. Baumgärtel Zur katalytischen Wasserdampfvergasung von Kohle bei hohen Drücken (Catalytic steam gasification of coal at high pressures) Chem.-Ing.Techn. 48 (1976) No 9, p 742

(3) F. Sabel Zur Frage der Gasbildung in Gaserzeugern (The formation of gas in gas producers) Chem.-Ing.Techn. 24 (1952), p 339

(4) K. Hedden
Die Bedeutung der Reaktionsgeschwindigkeit für die Abgaszusammensetzung bei Vergasungsprozessen (The influence of reaction velocity on exit gas composition in gasification processes)
Chem.-Ing.Techn. 30 (1958), p 125

3. The various routes to coal gasification

The chemical reactions involved in the gasification of coal are all heterogeneous gas-solid reactions accompanied by relatively large energy changes. Questions of material and heat transfer therefore play a prominent part in the design of coal gasification processes. It is an extremely difficult problem in process engineering to adjust the course of the chemical reactions on the one hand and the associated heat and material transfers on the other to the properties of the feed coal and to the construction materials and the stationary and dynamic behaviour of the gasifier in such a way that the raw gas produced matches as far as possible the use for which it is intended, or can be brought to the required specification by relatively simple treatment or conversion. Because of the complexity of the overall process it is obvious that numerous compromises are possible, or may be imposed by external constraints. This is the reason for the large number of different gasification processes which have already been tested or are still under development.

To achieve complete gasification it is necessary to transfer large quantities of heat to the coal in order to provide the heat of reaction. This suggests the classification of coal gasification processes on the basis of the technique of heat transfer chosen; each group will then contain processes which are similar as regards their energy utilization, their range of application and the composition of the gas produced. As a rough criterion we can distinguish between allothermic and autothermic processes, ie between those in which the heat of reaction is transferred to the reactor from outside, and those in which it is produced in the reactor itself by the gasification reaction.

In allothermic processes there is a further choice in

that either gaseous, liquid or solid heat carriers may be brought into direct contact with the coal to be heated, or the feed coal may be heated indirectly via heat exchange surfaces. All these variants have already been tested. Direct heat transfer is in fact used on a large industrial scale in the group of hot-blast gas processes, but owing to the low heat capacity of gases it is suitable in this form only for the gasification of lignites.

Solid heat carriers have been tested in some recently developed processes. As a rule the coal itself is used as the heat carrier: it is heated by blowing, which causes partial combustion, and then fed into the gasification reactor. The cyclic water-gas producer, in which blowing and gas production alternate (blow-or-make) and which at one time was used industrially on a large scale, may be regarded as a special case of a reactor with a solid heat carrier. Here, the exothermic and endothermic processes which in other cases are performed in different places are carried out in the same reactor.

Processes have also been proposed in which heat transfer is effected by liquids such as molten slag, salts or metals. As a rule, in order to utilize the high heat capacity of the liquid gasification and combustion are carried out autothermically in the same reaction space. The liquid transfer agent then serves to heat up the feed coal rapidly, and thus to even out short-term temperature fluctuations caused by irregularities in the feed rate of coal and gasifying agent. However, a two-chamber process has also been tested, in which combustion and gasification take place in separate compartments. Heat transfer between the two chambers is effected by molten slag.

In recent German developments of the steam gasification of coal employing process heat from high-temperature nuclear reactors the good heat transfer properties of fluidized beds are used to feed the heat removed from the nuclear reactor by means of helium indirectly into the gasifier via immersed heating tubes.

In autothermic processes the composition of the gasifying medium is generally the means by which the interplay between individual exothermic and endothermic reactions is adjusted in such a way that the overall process takes place in the manner required. Heat is most commonly supplied by the partial combustion of the coal with air or oxygen. As a result the raw gas will contain (apart from any nitrogen from the air) an appreciably higher content of $CO + CO_2$ than is the case with allothermic processes. On the other hand, autothermic processes possess the great advantage of avoiding any losses associated with heat transfer. The construction of the gasifier is also simplified. Further, the amount of heat needed for the reaction can be adjusted quickly and accurately as required by regulating the oxygen content of the gasifying agent. On account of these advantages, all the gasification processes which are at present being operated on a large industrial scale (cf Part A) are without exception autothermic processes.

In particular, the use of oxygen makes it possible to run autothermic gasification generators at a considerably higher temperature than can be achieved with allothermic gasifiers. This markedly accelerates the reactions involved in gasification. As a result the throughput and the decomposition of the steam are considerably increased. This makes it possible to gasify relatively inert coals, or coals with a high ash content. However, in adjusting the temperature care must be taken to ensure that the ash does not soften, and can be removed dry: alternatively, in modified generators with refractory linings the ash can be removed as slag. Operation anywhere between these two extremes causes the ash to cake into a soft dough-like mass in the gasifier, thus interrupting operation. The conditions are regulated by adjusting the steam/O_2 ratio.

However, partial combustion is not the only side-reaction which can be used to supply heat. In all gasifying reactors operating at a sufficiently low temperature the hydrogen produced from steam reacts to a certain extent to give methane (Section 2, reaction e). This reaction is exothermic and can contribute significantly to the heat balance in the reactor. In hydrogasification, however, in which the coal is treated directly with hydrogen as the gasifying medium, it is necessary to heat the reactants, and there always remains an ungasified residue containing carbon which must be caused to react by some other means if a satisfactory carbon utilization rate is to be attained.

A smaller contribution to the heat balance is made by the less strongly exothermic water gas shift reaction (Section 2, reaction f), which also occurs in the gasifier between excess steam and carbon monoxide when the temperature is not too high.

In the American CO_2-acceptor process, the exothermic reaction of CO_2 formed during gasification with calcined dolomite added to the coal is used to provide at least part of the heat needed for gasification by steam. However, since this reaction does not go to completion and cannot supply sufficient heat, the acceptor and the coal have to be preheated. Both this process and hydrogasification can therefore be described as combined auto-allothermic processes.

Although heat transfer is the essential step in coal gasification processes inasmuch as they involve energy changes, in practice the conveyance of the coal and the gasifying medium is usually the crucial aspect to which the question of heat transfer has to be subordinated. It seemed useful, therefore, to classify gasification processes on the basis of the mode of conveyance chosen.

Counter-current flow of the coal and the gasifying

agent, as used in most fixed-bed generators, causes the reactions involved in gasification to dominate at different points and at different temperatures in the bed of coal. The fresh coal is preheated by the hydrogen-containing gas and is carbonized and hydrogenated, so that the raw gas contains relatively large amounts of tars, oils and gaseous hydrocarbons. The carbonization char is largely consumed by the very hot steam in the gasification zone of the generator, while the most unreactive carbon residues are almost entirely burned in a lower combustion zone with the oxygen of the blast gas. The gasifying agent is heated up as it passes through the bed of hot ash and the combustion zone, and the raw gas produced is cooled down again in the carbonizing and preheating zones, so that in addition to very good conveyance of material an excellent heat utilization is obtained. Gasification in fixed-bed generators is of course only suitable for lump coals which form a sufficiently loose-packed mass. For a long time fixed-bed reactors were restricted to coal or coke which is non-caking, or at most weakly caking. However, recent developments with the pressurized Lurgi gasifier have rendered it suitable also for use with caking coals.

A great disadvantage of the earlier fixed-bed generators lay in their relatively low throughput, which was also unsatisfactory for very large-size coal. Two different solutions were adopted: firstly, raising the temperature above the melting point of the ash (slagging-bed generators), and secondly, operating under pressure. Both alternatives have been operated successfully. Slagging-bed generators of this type are, however, suitable only for gasifying coke: they are no longer in use today. On the other hand, pressure gasifiers, which are suitable for all lump coals because they incorporate stirring arms, are still in large-scale industrial operation nowadays. Attempts are currently being made to bring a pressurized slagging generator into industrial use. This, too, requires a lump coal which is at most weakly caking. The comparatively high throughputs achieved in this gasifier are limited by the capacity of the coal lockhoppers.

If a current of air is passed through a bed of fine-grained material at increased blast pressure to achieve a high enough flow velocity, the solid particles separate from one another and are set in turbulent circulatory motion. This principle has also been used in the gasification of coal. In a fluidized-bed reactor no reaction zones are able to form as they do in a fixed bed, owing to the intensive internal mixing of the solid particles. Instead, the fluidized state of the solid results in a very homogeneous temperature distribution throughout the whole volume of this turbulent layer, so that the gasification reactions take place much more uniformly. In contrast to the behaviour of the solid, however, the composition of the gas varies continuously through the fluidized bed, since combustion processes predominate in the lower part of the bed, and gasification in the upper part. Carbonization no longer occurs separately. Instead, the feed coal supplied is heated very rapidly to the reaction temperature.

Any volatile components released are cracked by hydrogenation by the hydrogen-containing raw gas, which leaves the fluidized bed without being cooled down, so that at the gasifier exit the raw gas contains hardly any tar or hydrocarbons. As a result of the grading effect of the gas stream a larger proportion of fine particles is inevitably carried out, and the dust content of the raw gas is therefore very high. This dust generally still has a high content of carbon, and on account of its large internal surface can be used as a filtering agent: alternatively, it can be returned to the reactor.

Owing to the turbulent motion of the solid particles in the fluidized bed, which have a high heat capacity, heat transfer from the bed to the walls or to inserted heating surfaces is extremely efficient. For this reason a recent process seeks to introduce the heat of gasification into the gas generator allothermically via a heat exchanger ("immersion heater process"). For example, the heat of gasification can be introduced "neat" into the gasifier from a high-temperature nuclear reactor via the reactor coolant, and need not entail burning any fossil fuel. In this way the yield of gas, expressed in relation to the amount of coal used, is considerably improved and the emission of CO_2 correspondingly reduced.

Because of the special flow properties of the turbulent layer, the fluidized bed technique is adaptable to varying loads without any serious effect on its thermal efficiency. This advantage, coupled with the very homogeneous temperature distribution, is the reason why the fluidized bed technique is used in most of the new gasification processes. However, it suffers from two disadvantages: the temperature has to be kept below the softening point of the ash, and the dimensions of the gasifier need to be large in order to include a gas stabilization chamber above the reaction zone. In more recently developed processes these disadvantages have been countered on the one hand by adding "anti-flux" agents and, on the other, by raising the pressure.

Because of the intensive internal mixing of the solid material, fluidized-bed gasifiers are not counter-current devices, but have properties far more closely resembling those of cross-current or co-current reactors. For example, the exit gas leaves at the reaction temperature of the fluidized bed. For this reason particularly good waste heat recovery is needed in order to obtain a high thermal efficiency.

If the flow velocity of the gasifying medium is increased so much that the tractive force exerted on the solid particles exceeds their weight, the region of pneumatic transport is attained (entrainment). By raising the reaction temperature, it is also possible to carry out coal gasification in entrained flow while keeping the reactor volume within reasonable limits, since the extremely high velocity of the chemical reactions means that the residence time in the reaction zone can be kept sufficiently short. In entrainment gasifiers of this kind the reaction zone is merely a flame, through which the coal and the gasifying medium pass together. In order to obtain complete utilization of the coal and satisfactory slag run-off, high reaction temperatures (above 1500°C) are needed, and these are obtained by using a gasifying medium containing only a small proportion of steam. Gasification then occurs predominantly by partial oxidation of the carbon with a sub-stoichiometric amount of oxygen. Gasification by steam plays only a secondary role, serving mainly to keep the temperature below an upper limit determined by the material of the reactor.

The outcome of this procedure, depending on the reactivity of the feed coal, is a raw gas with a high content of carbon monoxide which must in general be converted before it can be used further. As regards its use as synthesis gas, it is obviously an advantage that the gas contains only traces of methane and is free from tar and other higher hydrocarbons.

Entrained-flow reactors can use any type of coal, but it must be very finely ground in order to obtain a high degree of carbon gasification. Owing to the high temperature of the exit gas a sophisticated heat recovery system is necessary in order to obtain a good thermal efficiency. Because of the need to reach high reaction temperatures the oxygen consumption of this gasifier is relatively high compared with that of other types. The co-current flow of coal and gasifying medium leads to unfavourable conditions for heat and material transfer. In some entrained-flow processes the reactor is therefore constructed as a vortex chamber with tangential injection. The difference in inertia between the solid and the gas means that the two phases move relative to one another in the reaction zone, and material transfer is improved.

The operation of entrained-flow gasifiers under pressure should yield high output densities. However, the problems of feeding coal dust into a pressurized gasifier and of regulating its injection into the reaction zone have not yet been satisfactorily solved, and are at present the major difficulties in these new developments.

In gasifiers with liquid heat carriers the coal and the gasifying medium are blown on to or into a bath of the molten heat transfer material, which owing to its high heat capacity ensures uniform and stable reaction even with large and rapid fluctuations in load. Coal slag and molten salts and metals have been suggested as heat carriers. The advantages of this method are the relatively simple removal of residues, since both the residue and the heat carrier are in the liquid state, and the ease with which additives to improve the fluidity of the slag can be introduced. Some additives can also at this stage largely desulphurize the raw gas, which in any case is free from tars and higher hydrocarbons because of the high reaction temperature. The co-current flow of coal and gasifying medium is favourably influenced by the recycling of heat in the heat transfer layer, but the oxygen consumption of these generators too is comparatively high. Sophisticated waste heat recovery is necessary to ensure economic operation.

Fuel gases obtained by coal gasification can be classified according to their calorific value into three main groups, which also correspond roughly to their range of application. The term lean gas is used to describe gases obtained by gasification with air and steam which have a calorific value of about

3.8-7.6 MJ/m^3 (iN). Because of their low specific energy content lean gases are unsuited for transport, and instead are used preferably for firing purposes in the immediate vicinity of their production. Lean gas production is particularly important as a stage in the operation of coal-fired power stations using gas/steam turbines.

Gasification with oxygen and steam yields gases with calorific values in the approximate range of 10.5-16 MJ/m^3 (iN), which can be treated to give hydrogen, reducing gas or synthesis gas, or can be used as fuel gases either directly or after carburation.

Hydrogasification of coal gives gases which are particularly rich in methane, having calorific values greater than 21 MJ/m^3 (iN): they are particularly suited to conversion to SNG (substitute natural gas).

References

Summaries and reviews:

(1) H.W.v. Gratkowski
in Ullmanns Encyklopädie der techn. Chemie (Ullmann's Encyclopaedia of Industrial Chemistry) 3rd edition, Vol 10, p 396, Munich and Berlin 1958

(2) J. Schmidt
Technologie der Gaserzeugung (Technology of gas production), Vol II: Gasification
Leipzig 1966

(3) M. Meunier
Vergasung fester Brennstoffe und oxydative Umwandlung von Kohlenwasserstoffen (Gasification of solid fuels and oxidative conversion of hydrocarbons)
Weinheim 1962

(4) I. Howard-Smith and G.J. Werner
Coal conversion technology. A review
New Jersey 1976

(5) W. Osterle
Übersicht über die Verfahren zur Kohlevergasung (Review of coal gasification processes)
Battelle Institut, Frankfurt 1975

(6) R.A. Glenn (Ed)
Gas generator research and development
Survey and evaluation, 2 vols
Bituminous Coal Res., Inc, Monroeville 1965
BCR Report L-156 for OCR, Dept of Int, Washington DC

(7) K. Littlewood
Gasification: theory and application
Prog. Energy Combust. Sci. 3 (1977), pp 35-71

4. Main lines of development in new process design

As already mentioned in the Introduction, future prospects for the application of gas-from-coal technology lie in the following fields:

a. production of SNG (substitute natural gas);

b. production of synthesis gas for the chemical industry;

c. production of fuel gases for industry and for power stations;

d. production of reducing gas for metallurgical purposes, eg the direct reduction of iron ores.

Development projects have therefore been guided by the requirements to be met by gas used for one of the above purposes.

For all these applications it is advantageous to have the gas available under pressure. The modification of existing processes for pressurized operation or the design of new pressure processes therefore play the leading part in present-day development programmes. In the first place raising the pressure increases the throughput, since the concentration of the gasifying medium is correspondingly increased. At the same time the gasification equilibria are also affected, as described in Section 2. For example, the proportion of methane in the raw gas is increased, which is a great advantage in the production of SNG but is a disadvantage in other cases since it necessitates an additional separation or conversion (reforming). On the other hand, the methane content can be kept sufficiently low by raising the temperature, thus also making it possible, for example, to produce synthesis gas advantageously under pressure. An increase in pressure also has a favourable effect on the scale of the plant needed for gasifiers and gas scrubbers, and also allows considerable savings in the cost of compression, since the gas produced has in general a greater volume than the gasifying medium. Furthermore, operation under pressure makes it possible to purify the gas by physical methods which are impossible or uneconomic at atmospheric pressure and which carry the advantages of low costs for the scrubbing liquid, minimal consumption of scrubbing liquid and a generally lower energy consumption for its regeneration.

Another primary aim in the development of many gasification processes is an increase in the reaction temperature. This serves, above all, to increase the throughput and hence to reduce the specific dimensions of the gasifier. In addition, the production of tars and unwanted hydrocarbons is decreased. The displacement of equilibria at high temperatures leads to a lower content of CO_2 and more complete decomposition of steam (cf Figures 1.1 and 1.2), thus producing a gas with a higher reducing potential.

Gas used for the direct reduction of iron ores must have a reducing potential of more than 90% (reducing potential = vol % of the reducing components $CO + H_2$). With some processes it is hoped to satisfy this requirement without further treatment of the gas.

All fluidized-bed processes, which feature very prominently among recent developments, are restricted to a maximum temperature fixed by the melting point of the ash, since if the ash melts the fluidized bed will collapse. This limit does not apply to fixed-bed, slagging and entrained-flow processes, in which the ash can be removed in a molten state. In these cases temperatures of as much as 1500-1900°C can be used in the gasifier: indeed, these very high temperatures are needed to produce a good slag flow. The temperature range in which the ash softens but does not become molten (ca 1000-1500°C) can, on the contrary, be used only with great difficulty for coal gasification.

In high-temperature gasifiers the removal of the slag in a molten state and the generally higher temperature of the raw gas lead to heat losses which are greater than those for reactors with dry ash removal. Radiation losses are also higher, so that overall the oxygen consumption is substantially greater. This disadvantage is partly cancelled out by the considerably lower consumption of steam. This is partly because much less steam is required to regulate the upper temperature limit than is the case in fixed-bed reactors with dry ash removal, and partly because the steam is almost completely decomposed at the high temperatures involved, so that its cooling effect is greater. The lower steam consumption also reduces the amount of gas liquor, which is a further advantage. This represents a considerable gain, since when added to the absence of oils and tars it considerably simplifies the anti-pollution measures required with high-temperature gasifiers: with fixed-bed processes, for example, treatment of the gas liquor is comparatively expensive because of its content of non-biodegradable substances.

For the production of gases of high calorific value (SNG) it is advantageous from the energy point of view for the raw gas itself to contain a large proportion of methane, since in the methanation reaction (reaction g) a large proportion of the chemical energy of the reactants is liberated in the form of heat. However, the use of higher temperatures suppresses the formation of methane (cf Figure 1.3). A number of new processes therefore seek to carry out the individual gasification reactions in different temperature ranges by separating the process into several stages, and thus to some extent to optimize separately reactions involving different products. For example, in the American Hygas process the coal is first dried and then cracked by hydrogenation in suspension at 650°C, using a hydrogen-rich gas which is obtained by two-stage steam gasification of the char in fluidized beds at 950°C and 1050°C. In this way raw gas containing 18% methane has been obtained.

A totally different line of development seeks to dispense with the costly oxygen plant. The oxygen needed for gasification is usually obtained by the fractional distillation of liquid air at low temperatures. This process involves high capital and running costs, which make the gas produced more expensive. Some processes therefore seek to separate the steam gasification and the heating of the coal, so that air can be used instead of pure oxygen for combustion. The heat of gasification can be supplied to the reactor allothermically by special heat carriers or heat exchangers, or alternatively the coal itself can be heated by partial combustion in a separate heater before it enters the reactor. Gasification processes using heat from nuclear reactors also belong to this group. This involves going a step further, of course, in that no coal is burned to provide the heat of reaction, but nuclear fission in a high-temperature nuclear reactor serves as the source of heat. Such a procedure dispenses with the oxygen plant and also saves part of the coal feed. This type of process naturally presupposes that a suitable source of high-temperature heat is available at a favourable cost. This source can still be used in the low-temperature range for producing electricity.

These observations show that separation of the oxidation and gasification reactions automatically stipulates an allothermic procedure. The advantages lie in the elimination of the oxygen plant and the separation of heat production (CO_2) and gas production (water gas), while the disadvantages are greater losses through heat transfer and a narrowing of the range of operating temperatures. These must be weighed against one another in each individual case, and no general preference is possible.

In developing the processes described above the aim is to produce gas with the highest possible calorific value. In processes which produce gas for direct firing, for example in power stations, the object is, however, simply to transfer the energy content of the coal temporarily to a gaseous carrier which can be more readily handled and used in a gas/steam turbine. Since transport involves only short distances and no storage is necessary, the nitrogen content causes no difficulties in this case and air can in general be used as the gasifying medium. For specific use in combined-cycle gas/steam turbine processes it is particularly important that the gas should be under pressure. Most such combined processes now under development also seek to exploit the adaptability of fluidized-bed generators to large load fluctuations. There are, however, problems connected with gas purification. Because of the high nitrogen content the quantities of raw gas are large, and carry out much dust with them from the gasifier, so that gas scrubbers need to be redesigned. In addition, all scrubbing processes have to be carried out above the steam dew-point, so as to minimize mass losses in the raw gas. For this purpose a number of new scrubbing processes have been devised and are already in the course of industrial development. It would be ideal (and indeed appears to be of paramount importance for the purification of reducing gas for metallurgical purposes) if the removal of dust and sulphur could be carried out at a high temperature, close to that at which the gas leaves the gasifier, so that gasification and subsequent utilization can be carried "in one run".

Another important aspect of development concerns the improvement of existing mechanical equipment. For instance, in fixed-bed processes attempts are being made to permit the use of strongly caking coals by building in suitable agitators. In all pressure processes the problem of feeding the coal into the pressure reactor has claimed priority in development programmes. In conventional Lurgi pressure gasifiers pressure lockhoppers are sufficient, but in high-temperature pressure gasifiers, which have a considerably greater throughput, new systems of injection are already needed, since the throughput is at present limited by the rate at which coal can be fed in. The extremely large fluxes of solid material which have to be handled in modern high-duty processes are at present the limiting factor for any further increase in the performance of coal gasification processes. For the immediate future, further development of the techniques of mechanical handling appears to take precedence over solution of the thermal problems of coal gasification.

5. Explanation of the form of presentation

Since there are so many parameters which characterize a coal gasification process, the classification of processes for the purposes of presentation could be based on various different principles. The weight attached to individual factors is to a large extent arbitrary, and will depend on the aim in view. The review of the subject given in this book is intended to provide a survey of the present position as regards technological prospects. For this reason the classification adopted has been based mainly on the industrial development status of a process and its underlying principles. More specifically, a distinction is made between processes which are currently in commercial operation, gasification processes which are no longer operated or which have not yet come into operation, and processes which are still at the planning or laboratory investigation stage.

Since they are totally new in concept, processes which use nuclear process heat have been described in summary in a separate chapter. This is also justified by the fact that these processes have been conceived on a much broader basis of association with large-scale installations (nuclear reactors), in order to produce heat, gas and electricity in an ecologically clean form.

In recent years there has been a revival of interest in the underground gasification of coal in situ. This is also a gasification process, and one which is today already being pursued in several countries, in some cases at considerable financial outlay. It therefore seems appropriate that any review of coal gasification should also contain some account of underground gasification. However, since the special features of this technique preclude any technical comparison with the above-ground processes described earlier, a few typical procedures have been grouped together in a separate chapter at the end of the book.

Within each group processes have been arranged according to the type of reactor used, starting with fixed-bed generators. These are followed by fluidized-bed processes, then entrained-flow reactor processes and, lastly, processes with liquid heat carriers.

A Processes currently in commercial operation (including further developments)

1. Revolving-grate generator processes (Two-shaft gas generators)
2. Lurgi process (Ruhr 100)
3. Winkler process (HTW, Davy Power Gas)
4. Koppers-Totzek process (Shell-Koppers)

B Processes not currently in commercial operation and processes under development

1. Hot-blast gas processes
2. Slagging-bed generator processes
3. British Gas Corporation process
4. Hygas process
5. Cogas process
6. CO_2-acceptor process
7. Bi-Gas process
8. Ash-agglomerating process
9. Synthane process
10. U-gas process
11. Texaco process
12. Rummel-Otto process (Saarberg-Otto process)
13. Ruhrgas vortex chamber process

14. Combustion Engineering process
15. Westinghouse process

C Gasification processes still at the planning stage (small-scale experiments)

1. Exxon process
2. Hydrane process
3. Kellogg process
4. Atgas process

D Gasification of coal using nuclear process heat

1. Steam gasification of lignite
2. Steam gasification of hard coal
3. Hydrogasification of lignite
4. Hydrogasification of hard coal

E Examples of the use of underground gasification

Operational data for the Angren and Yuzhno-Abinsk installations (both USSR)

and results of experiments at Hanna (USA)

The review given here is concerned purely with gasification processes proper, and does not cover the various combined-cycle processes which are designed for the use of the product gas in a combined gas/steam turbine process. This decision is based on the fact that this step in the utilization of the gas can in principle be combined with any gasification process, though preferably with those which operate under pressure. Combined-cycle processes are therefore not tied to any particular process, though they have usually been developed in connection with a specific gasification technique.

Apart from the examples given under Group E, for all other processes special emphasis has been placed on the reactor in which the coal is converted. Only a brief description is given of the other units required in a gasification plant, for example for gas purification or methanation. The review is also confined to essentially typical gasification processes. Variants which do not differ from these processes in their essentials are not described separately.

As an aid to rapid assessment and comparison, the individual processes have been examined according to a uniform scheme, in which the items listed below (with the exception of the last) have been treated as completely and exhaustively as possible. The references to the literature have been chosen so as to provide further information about the details of each process. However, a complete coverage of the source material or the patent literature would go far beyond the intended scope of the book.

I	Gasification system
II	Performance
III	Feed coal requirements
IV	Raw gas composition and calorific value
V	Description of process
VI	Production of gasifying medium
VII	Technical data
VIII	Construction and development
	Literature references

Under heading I a brief account is given of the type of reactor, the gasifying medium and the operation of the process. Heading II gives data applicable to the size of reactor unit so far tested. It includes the throughput of coal per hour, the hourly production of gas and, where available, the dimensions of the reactor. All data given in the subsequent description of the process refer to reactors of this size unless otherwise stated.

The coal which can be used for the process in question is specified under heading III, as regards (a) rank, (b) particle size, (c) ash content and melting behaviour and (d) any necessary pretreatment of the coal, such as grinding and drying. The raw gas analysis given under heading IV always refers to the dry, tar-free gas at the reactor exit.

The description of the process (heading V) describes in as much detail as possible the passage of material through the reactor and any auxiliary units directly connected to it upstream and downstream, such as those for coal pretreatment, ash removal, heat recovery, and purification of the raw gas. In many cases further stages of gas purification have also been included if they bear a special relation to the gasification process. This heading also describes the dynamic behaviour of the gasification reactor where such information is available, but data on this have been published in only a few instances.

In describing the production of the gasifying medium (heading VI), a special effort has been made to state whether the waste steam which is usually produced can be used for gasification, or whether steam has to be obtained from other installations.

Numerical values for the most important technical characteristics of each process have been grouped under heading VII, which should aid direct comparison between different processes. In many instances the necessary figures were not available in this form in the literature, but wherever an unambiguous calculation or a simple recalculation was possible, this has been done. Estimates have been used only in exceptional cases, and then only when they could be carried out simply and with reasonable certainty.

In many cases it is possible to specify efficiencies for the gasification processes described. The following definitions have been used throughout:

Carbon conversion efficiency:

ratio of the amount of carbon contained in the products obtained by gasification to the amount of carbon contained in the fuel;

Gasification efficiency:

ratio of the amount of chemically bound heat in the dry gas to the amount of chemically bound heat in the fuel;

Thermal efficiency:

ratio of the amount of useful chemically bound and sensible heat in the products obtained by gasification to the amount of chemically bound and sensible heat in the fuel and the gasifying medium.

The specific data refer to the coal consumed, the amount of raw gas produced, and the energy in GJ (Gigajoules) in the raw gas produced (having the composition and calorific value given under heading IV). The exit gas temperatures given are those immediately at the outlet of the reactor, ie before the gas enters the waste heat boiler.

For the majority of processes there are no published figures available on the water and energy consumption of gasification plants.

Heading VIII lists the developers, manufacturers and operators of the processes described where these are known.

The flow-sheets and sectional drawings of the reactors have been taken from the most recent literature, or constructed from published data. The Sankey diagrams have mostly been specially constructed for the book from the data reported, since they were available only in a very few cases.

6. Survey of gasification processes

The following tables list the most important of the coal gasification processes dealt with in the present book, together with some characteristic data which provide at least a brief survey of the technological status of these processes. However, since a meaningful description of a coal gasification process requires a large number of parameters which cannot be presented in tabular form, these tables are necessarily of limited value and constitute no more than a rough guide. It is in the later sections giving a detailed description of each individual process that an attempt has been made to assemble and present the relevant parameters as adequately as possible. To arrive at a valid assessment, these detailed descriptions should therefore always be consulted.

The tables set out the following brief characteristics for each process: Row 1 indicates the industrial development status of the process. Rows 2, 3 and 4 show the requirements as regards the feed coal and the nature of the gasifying medium, as far as these have been published. For some processes, however, successful experiments have already been carried out to extend the range of coals which can be used, although they involve modifying the design of the reactor (eg the Lurgi pressure gasification process).

Row 5 states the method used for conveying the fuel and the gasifying medium. Row 6 contains data on the operating pressure. The next three rows give information on the process products, ie row 7 (raw gas composition and calorific value), row 8 (specific yield of gas) and row 9 (by-products). The calorific value of the raw gas determines its range of applications. A calorific value of less than 7 MJ/m^3 (iN) (raw gas, dry basis) is obtained by partial combustion with air, and represents a heating or fuel gas with a high content of nitrogen. Partial combustion with oxygen gives a water gas with calorific values of up to about 11.5 MJ/m^3 (iN) (raw gas, dry basis), which can be used as synthesis gas, etc. Raw gas of higher calorific value than this usually contains appreciable amounts of methane, which makes it particularly suitable for transport over long distances or for further treatment to give SNG. In the case of hydrogasification, calorific values of more than 20 MJ/m^3 (iN) (raw gas, dry basis) can be obtained. This procedure leaves some residual char, but this can be used, for example, to produce hydrogen.

Rows 10 and 11 indicate the average specific consumption of the gasifying agents (oxygen and steam). The last three rows (rows 12, 13 and 14) give the three efficiencies defined earlier in Section 5, so far as these can be obtained from the literature.

Tabular survey of the major coal gasification processes I

	1	2	3	4	5	6
Process	Lurgi	Winkler	Cogas	CO_2-acceptor	Synthane	Koppers-Totzek
Development status	successful on a large industrial scale	successful on a large industrial scale	experimental plant in operation	experimental plant in operation	experimental plant in operation	successful on a large industrial scale
Coal requirements	lignites, weakly caking coals	mainly lignites	lignites, young coals	lignites	all non-anthracitic coals	any
Coal size	lump coal, 6-40 mm	fine particles, < 8 mm	no data	0.15-2.4 mm	dust, < 0.85 mm	dust, < 0.1 mm
Gasifying medium	O_2 + steam	O_2 + steam	steam	steam	O_2 + steam	O_2 + steam
Conveyance of material Type of reactor	counter-current, fixed-bed, revolving grate	fluidized bed	fluidized bed solid heat carrier	fluidized bed solid heat carrier	fluidized bed	co-current, entrained-flow
Pressure	20-30 bar	atmospheric	1.5-3 bar	11 bar	40-70 bar	atmospheric
Raw gas analysis			Mixed gas:	Mixed gas:		
H_2	36-40%	35-45%	58 %	56 %	28%	22-32%
CO	18-25%	30-50%	31 %	15.5%	17%	55-66%
CO_2	27-32%	13-25%	6.6%	11 %	29%	7-12%
CH_4	9-10%	0.5-2%	4 %	14 %	24%	0.1%
Calorific value MJ/m^3 (iN)	11.5	9-13	12.9	14.6	15.9	10.6-11.8
Specific yield of gas m^3(iN)/t of coal	1400-1700	1600	no data	1360	no data	1650
By-products	tar, oil, gas naphtha, ammonia	much dust, hardly any higher hydrocarbons	suspended dust, oil products	no data	no data	none
Specific O_2 consumption	220-300 m^3(iN)	350 m^3(iN)	not applicable	not applicable	no data	540 m^3(iN)
Specific steam consumption	1-1.4 t (30 bar)	0.88 t low-pressure steam	no data	3.6 t	no data	0.24 t low-pressure steam
Carbon conversion efficiency	99%	90%	96%	72%	60-70%	90-96%
Gasification efficiency	75-85%	≤ 75%	68%	77%	no data	70-77%
Thermal efficiency	75%	80%	86%	(> 90%)	no data	75-85%

Tabular survey of the major coal gasification processes II

	7	8	9	10	11	12	
Process	Texaco	Bi-Gas	Hygas	Rummel-Otto (Saarberg-Otto)	Processes using heat from high-temperature nuclear reactors: Steam gasification	Hydrogasification	
Development status	tested on a large industrial scale	experimental plant in operation	experimental plant in operation	tested on a large industrial scale	semi-scale plant in operation	semi-scale plant in operation	
Coal requirements	any	any	all non-anthracitic coals	any	hard coals	lignites	
Coal size	dust < 0.75 mm	dust < 1 mm	dust < 1 mm	fine particles < 3 mm	dust < 0.1 mm	dust < 1 mm	
Gasifying medium	O_2 + steam	O_2 + steam	O_2 + steam	O_2 + steam (air)	steam	hydrogen	
Conveyance of material Type of reactor	co-current entrained-flow	co-current entrained-flow, two-stage	combined multi-stage process, entrained-flow/ fluidized bed	co-current liquid heat carrier	fluidized bed	fluidized bed	
Pressure	15-85 bar	35-100 bar	70 bar	atmospheric	40 bar	80 bar	
Raw gas analysis				(gasifying medium O_2 + steam)	mixed gas	with methane reforming	with hydrogenation gas preheating
H_2	38%	32%	30%	40%	54%	63 %	38%
CO	46%	21%	24%	40-50%	19%	4 %	15%
CO_2	5-20%	29%	24.5%	10%	18%	0.7%	16%
CH_4	0.8%	15%	18.5%	traces	6%	31 %	21%
Calorific value MJ/m^3 (iN)	11.1	13.6	15	11	12.9	21	15
Specific yield of gas m^3(iN)/t of coal	1800-2200	no data	no data	1800-2000	2340	890	810
By-products	none	no data	liquid hydro-carbons	none	tar, oil, ammonia	residual char	residual char
Specific O_2 consumption	585 m^3(iN)	450 m^3(iN)	1370 m^3(iN)	580 m^3(iN)	not applicable	not applicable H_2:550 m^3 (iN)/t of coal	not applicable
Specific steam consumption	0.7 t high-pressure steam	1.2 t high-pressure steam	0.96 t high-pressure steam	1 t low-pressure steam	2.2 t (43 bar)	0.44 t (41 bar)	0.33 t (82 bar)
Carbon conversion efficiency	90-98%	no data	96-99%	99%	no data	63%	60%
Gasification efficiency	75-80%	no data	no data	≥80%	(110%)	no data	no data
Thermal efficiency	≤ 85%	no data	no data	83%	no data	89%	93%

A. Processes currently in commercial operation

1. Revolving-grate generator processes

Grate generators represent the earliest type of equipment used industrially for the gasification of solid fuels. Difficulties in ash removal did, however, frequently disrupt or halt operation with this type of generator. The introduction of the revolving-grate generator around the turn of the century constituted an important improvement in design from the point of view of continuous operation and hence industrial use. Revolving-grate generators became very widespread. The use of this type of generator for producing lean gas is attracting renewed interest today, especially in regions where there is no SNG distribution network and coal is available at low cost. Since all revolving-grate generators are constructed in essentially the same way, it seems justifiable to treat them collectively despite minor differences in design. We shall therefore take as our example the operating results for the coal gasification by the Wellmann-Galusha process in the form described earlier. Other processes, which have been used in Germany in particular (eg those developed by DEMAG, Dr C. Otto GmbH and H. Koppers GmbH), are covered by literature references.

A revolving-grate generator generally consists of a cylindrical shaft, which in some types narrows to a cone in its lower sections. The diameter of the shaft is usually between 1.5 and 3 m, though it may reach 5 m in some instances. Its walls are either constructed as a water jacket, or may be either completely or partly brick-lined. The shaft has a sturdy cover, which carries the charging equipment. The lower part of the generator, which consists of the revolving grate and the water seal, can be rotated about the axis of the shaft by suitable driving gear.

The construction so far described is fairly standard,

but special features are sometimes added, especially for caking coals. For example, the upper part of the gasifier may be fitted with an agitator, which serves to loosen coal which has caked together. Sometimes the grate is fixed and the walls of the shaft can be rotated. Generally speaking, however, the essential features are basically the same, though there may be differences in the degree of automation, mechanical reliability, charging arrangements, and adaptability to different grades of coal.

The generator is fed with coal from above by the charging equipment. The gasifying medium, air saturated with steam at a high temperature or an oxygen/steam mixture, is injected from below through the revolving grate. An even distribution of air is ensured by the special design of the grate. The counter-current flow of the gasifying medium and the coal favours heat conservation, and also creates zones of decreasing temperature moving upwards through the charge, in which different processes occur preferentially. In the top part of the shaft the freshly added coal is first dried by the counter current of hot raw gas. In the region immediately below this the dried coal is carbonized by the hot gasifying medium, with release of pyrolysis products. This region is followed by the gasification zone, in which the preheated coal reacts with the very hot gasifying medium in an endothermic reaction. The very unreactive residual char is burned completely in the combustion zone beneath, thus providing the heat needed for gasification. A hot layer of ash lies immediately above the grate, in which the gasifying medium is preheated. The ash is removed continuously by very slow rotation of the flat or conical revolving grate, which is water-cooled in many generators, and is often constructed eccentrically to improve ash removal. There are sometimes additional devices for breaking up the lumps of slag. The ash can either be removed through the water seal at the base of the reactor or, in the case of closed reactors, extracted in a dry state. These two procedures are termed wet and dry ash removal respectively.

Revolving-grate generators are particularly suited to the gasification of coke; in this case the gas produced is not enriched by products of carbonization. When coal is used as the feed, carbonization liberates volatile constituents. This increases the calorific value of the gas, but on the other hand the presence of carbonization products, particularly tar and phenols and also certain gases such as hydrogen sulphide and ammonia, necessitates expensive treatment of the raw gas.

In the type of generator described here the products of carbonization are present at low concentrations in a large volume of raw gas. In order to obtain a more concentrated carbonization gas, so-called two-shaft generators were introduced, which have a separate carbonization shaft. In these generators a large proportion of the clear gas obtained by gasification is drawn off directly behind the gasification zone. The fresh coal is then carbonized by only the remainder of the gas. This procedure gives a carbonization temperature which is lower than that in single-shaft revolving-grate generators. The effect of this is to reduce the amount of fine dust. The tar produced is of better quality. It is less contaminated, contains less pitch, and can be condensed more easily.

A further by-product is produced in generators furnished with a boiler jacket. Depending on the type of generator, this produces steam at low to medium pressures. Where low-pressure steam (up to 1.5 bar) is produced it is used mainly to saturate the air for gasification. Medium-pressure steam at about 10 bar is used partly for the same purpose, but the remainder is suitable for many other uses.

So-called water-gas generators are usually also revolving-grate generators. However, instead of continuous operation with steam-saturated air they operate intermittently with alternating periods of blowing and gasification (blow-or-make). In the first phase blowing with air is used to heat the generator by the combustion of coal. The incandescent coke is then treated with steam in the second phase. The endothermic reaction cools down the coke and produces the required water gas with high contents of carbon monoxide and hydrogen. Gasification is then followed by another period of air blowing so as to reheat the generator. These revolving-grate generators were operated close to atmospheric pressure and reached throughputs of up to 3.5 t coal/h (80 t/day). They were in widespread use at one time and were being built for industrial use at least up to the end of the 1950s, but they became less important with the general decline in gas-from-coal production. However, there are still a number of installations in operation throughout the world today, and the renaissance in coal gasification offers the opportunity of developing new processes based on revolving-grate generators.

References

1. General

P. Dolch, Wassergas, Chemie und Technik der Wassergasverfahren (Water gas: chemistry and technology of water gas processes), Leipzig 1936

W. Gumz, Gas producers and blast furnaces New York 1950

P. Dolch, Gas- und Wasserf. 41 (1932), p 807

G.W.C. Allan, J. Inst. Fuel 28 (1955) Sept, p 458

Conference Proceedings, Int. Conf. Complete Gasif. Mined Coal, Liège, 3-8 May 1954

Annales des Mines de Belgique, Brussels 1954

H. von Gratkowski in Ullmann's Encyklopädie der techn. Chemie (Encyclopaedia of Industrial Chemistry), 3rd edition, Vol 10, Munich and Berlin 1958

2. Particular processes

Kerpely synthesis gas process:

H.R. Batchelder, R.G. Dressler, R.F. Tenney, R.E. Krüger, R.D. Siegur
Am. Gas. Ass. Proc. 348-354 (1950)

L.F. Willmott, H.R. Batchelder, R.F. Tenney
US Bureau of Mines Rep. Invest. 5108 (1955)

Manufacturer: not stated
Operator: US Bureau of Mines
Location: Louisiana Missouri (USA)

Power Gas Mechanical process:

P.M.K. Embling, J.Inst.Fuel. 18 (101) (1945) 114
P.M.K. Embling, Int. Conf. Complete Gasif. Mined Coal, Liège 1954, p 66
Brit. Patent 539 664 (19 September 1941)
Brit. Patent 547 155 (17 August 1942)

Manufacturer: The Power Gas Corp, Ltd
Operator: not stated
Location: a. South Staffordshire (UK)
b. Oldbury (UK)

United Gas Improvement (UGI) process:

J.W. Blatchford, Am. Gas Ass. Progr. 652 (1950)

Manufacturer: United Engineers and Constructors
Operator: a. E.I. Du Pont de Nemours
b. Philadelphia Gasworks
Location: a. Belle, W. Virginia (USA)
b. Philadelphia (USA)

Kellogg process:

Anon, Oil and Gas Journal 72 (36) (1974) pp 86-88

Manufacturer: M.W. Kellogg Co
Operator: not stated
Location: not stated

PCV process (KGN process):

Booklet published by Projektierung Chemische Verfahrenstechnik GmbH (PCV), Düsseldorf Vergasung ballastreicher Steinkohlen (The gasification of coal with a high inert content)

Anon, works journal of Gewerkschaft Sophia-Jacoba 1977, No 2

Manufacturer: not stated
Operator: Kohlegas Nordrhein GmbH (joint enterprise of Gewerkschaft Sophia-Jacoba and PCV)
Location: Sophia-Jacoba colliery, Hückelhoven

Integral Engineering, Vienna is the owner of the VIAG process:

J. Williams, Industrial fuel gas from coal by two-stage gasification, booklet published by Woodall-Duckham Ltd, Crawley, UK, 1974

Schuster, Gas- u. Wasserfach 93 (1952) 57
F. Domann and H. Schmitt, Erdöl u. Kohle 12 (1959) 883

Manufacturer: Il Gas Integrale s.p.A, Milan, (subsidiary of Woodall-Duckham Ltd, Crawley, UK)
Operators: many municipal gasworks (about 100 units)
Location: formerly Rome, Strasbourg, Ulm, Freiburg, etc

International Furnace Equipment (IFE) process:

Anon, Modern Gas Plant, booklet published by International Furnace Equipment Co Ltd, Birmingham, UK

Manufacturer: International Furnace Equipment Co Ltd
Operator: not stated
Location: not stated

Demag process (Demag-P.-H. generator):

S.A. Burke, G.A. Sparham, J. Iron and Steel Inst. 172 (1952), p 257
O.V. Neudeck, Paper A 3, Intern. Conference on Coal Gasification, Liège, May 1954
Brennstoff Wärme Kraft 4 (1952), p 128

GE (General Electric) process:

Anon, World Coal 4 (1978) April, p 36

Manufacturer: not stated
Operator: General Electric Co
Location: Schenectady, New York, USA

The generators developed by Dr C. Otto u. Co GmbH and H. Koppers GmbH are described in Ullmann's Encyklopädie der techn. Chemie (Encyclopaedia of Industrial Chemistry), 3rd edition, Vol 10, Verlag Urban & Schwarzenbach, Munich-Berlin 1958

Wellmann-Galusha process (example of revolving-grate processes)

Coal gasification with air and steam

I. Gasification system

Autothermic gasification of coal with air and steam at atmospheric pressure in a stationary charge in a revolving-grate generator. The process was developed in the USA by the McDowell-Wellmann Engineering Company and used by McDowell in two commercial installations for producing a heating gas from coal. Numerous smaller installations have been built by other firms, but it is not known whether they are still in operation. According to the most recent reports Wellmann-Galusha gasifiers are again being brought into use in several places in the USA.

Operating data are given below for when air is used, followed by experimental results for the use of oxygen and steam.

II. Performance

The following figures represent throughput and production for a generator with an internal diameter of 3 m and an internal cross section of 7.3 m^2:

3.2 t coal/h

12 100 m^3(iN) gas/h

III. Feed coal requirements

a. Rank: The results given were obtained with a highly volatile coal (Logan County, Cedar Grove Seam, W Virginia). In principle both caking and non-caking coals are suitable, though an agitator is necessary for the former.

Coal analysis

Moisture	3.7 wt %
Volatile constituents	36.3 wt %
Carbon (fixed)	56.6 wt %
Ash	3.5 wt %
Calorific value	32.54 MJ/kg (= 7772 kcal/kg)

b. Particle size: 6-50 mm

c. Ash: Softening point 1450°C, ash content 3.5%. Ash is removed continuously via the eccentric revolving grate.

d. Coal pretreatment: not stated.

IV. Raw gas (dry, tar-free basis, measured at the generator exit)

Composition:	H_2	-	15.0 vol %
	CO	-	28.6 vol %
	CO_2	-	3.4 vol %
	CH_4	-	2.7 vol %
	N_2	-	50.3 vol %
	H_2S	-	not stated

Calorific value: 6.587 MJ/m^3(iN) (= 1573 kcal/m^3(iN))

V. Description

At the top of the generator the coal is fed from the storage bunker through several coal lockhoppers into the gasification space. An agitator ensures uniform distribution and prevents agglomeration. The gasifying medium (air saturated with steam) flows upwards from the base of the generator counter-current to the coal. The coal is gasified and the ash is removed continuously via the eccentrically mounted revolving grate into an ash gate lockhopper. The generator is cooled with water, and the steam thus produced serves as gasifying medium. Dust separation and gas purification are based on conventional techniques. No further details are available.

VI. Production of the gasifying medium

a. Air: air blast

b. Steam: saturation of the gasification blast with steam from the generator jacket

VII. Technical data

a.	Carbon conversion efficiency:	98%
b.	Gasification efficiency:	72%
c.	Thermal efficiency:	(92%)
d.	% steam decomposition:	25.5%

e. Inputs

Air:	10 300 m^3(iN)/h
	0.85 m^3(iN)/m^3(iN) raw gas
	3 200 m^3(iN)/t coal
Steam:	small quantity; not stated
Coal:	3.2 t/h
	0.264 kg/m^3(iN) raw gas
	0.04 t/GJ

f. Steam/oxygen ratio: not applicable

g. Products: gas yield: 3788 m^3(iN)raw gas/t coal
tar: 0.1 kg/m^3(iN) raw gas

h. Pressure: atmospheric pressure (sometimes a slight excess pressure, 1.1-3 bar)

Temperature:	not stated
Exit gas temperature:	600-650°C
Exit gas pressure:	ca 1.15 bar

i. Water consumption for generator cooling:

5.5 m^3/t coal
17.5 m^3/h
0.22 m^3/GJ
1.5 l/m^3(iN) raw gas

j. Energy and coal consumption: gasification: 3.2 t coal/h
energy: 50 kWh/t coal

k. Gas purification: because it contains suspended material the gas is of only limited use without further purification. Continuous operation over long periods is not possible, since the pipework needs to be cleaned every 3 or 4 months. Gas purification and cooling for further treatment to give a clean fuel gas add considerably to the capital and running costs.

VIII. Construction and development

McDowell-Wellman Engineering Co, Cleveland

Operator

McDowell-Wellman Engineering Co, Cleveland

References

(1) McDowell-Wellman, Bulletin 147, pp 1-16

(2) G.M. Hamilton in Annual Meeting AIME, St Louis, 1961

(3) G.M. Hamilton, Const. Eng. 8 (3) 4-11 (1963)

(4) Anon, Coal Age, May 1977, pp 37 and 79

Gasification with oxygen and steam

I. Gasification system

As before, the higher reaction temperature produced by operating with oxygen requires some modification of the reactor cooling jacket. The reactors are, however, essentially the same.

II. Performance

A generator with an internal diameter of 3 m, a combustion surface of 8 m^2 and a reactor volume of 21.7 m^3 gave the following operating results with coals A and B (see III):

	(A)	(B)	
gas production:	3200	2750	m^3(iN)/h
or	8.2	6.7	Gcal/h
coal throughput:	1.25	1.13	t/h

III. Feed coal requirements

a. Rank:

(A) Pennsylvania coal, St Nicholas Rice anthracite
calorific value: 28.6 MJ/kg (6800 kcal/kg)
dry: 10.9% ash (0.6% S), 4.8% volatile constituents, 84.3% carbon; swelling index 0

(B) Coke (dry): 12.5% ash (0.6% S), 87.5% carbon, swelling index 0, calorific value 27.9 MJ/kg (6666 kcal/kg)

b. Particle size:

(A) fines (< 10 mm)

(B) 2-70 mm

c. Ash

Softening point: (A) 1550°C
(B) 1200°C

The ash is removed continuously.

d. Coal pretreatment: drying, crushing, coking.

IV. Raw gas (dry, tar-free basis, measured at the generator exit)

Composition:	(A)	(B)
H_2	41.0 vol %	31.0 vol %
O_2 x)	0.1 vol %	0.6 vol %
CO	40.0 vol %	54.1 vol %
CO_2	16.5 vol %	11.3 vol %
CH_4	0.9 vol %	0.4 vol %
N_2	1.5 vol %	2.6 vol %
H_2S	not stated	not stated
Calorific value:	10.635 (2540)	10.132 MJ/m^3(iN) (2420 $kcal/m^3$(iN))

x) possibly analytical errors

V. Description

As before.

VI. Production of gasifying medium

a. O_2: oxygen plant

b. Steam: not stated

VII. Technical data

		(A)	(B)	
a.	Carbon conversion efficiency:	not stated	not stated	
b.	Gasification efficiency:	85.6%	81.9%	
c.	Thermal efficiency:	(95.6%)	88.8%	
d.	% steam decomposition:	53%	62%	
e.	Inputs			
	Oxygen:	540	558	m^3(iN)/t coal
		0.21	0.23	m^3(iN)/m^3(iN) raw gas
		20	23.4	m^3(iN)/GJ
	Steam:	1.41	1.0	t/t coal
		0.55	0.41	kg/m^3(iN) raw gas
		0.05	0.04	t/GJ
	Coal:	0.39	0.41	kg/m^3(iN) raw gas
		0.036	0.04	t/GJ
f.	Steam/oxygen ratio:	2.61	1.78	kg/m^3(iN)
g.	Products			
	Gas yield:	2564	2439	m^3(iN)raw gas/t coal
	Tar:	not stated	not stated	

h. Operating pressure: 1 bar

Operating temperature and exit gas temperature: not stated

i. Water consumption: not stated

j. Energy and coal consumption: not stated

k. Gas purification: not stated

VIII. Construction and development

As before.

Operator see above

Location see above

References

(1) C.C. Wright, L.L. Newman, Am. Gas Assoc. Proc. (1947), 701

(2) C.C. Wright, K.M. Borchay, R.F. Mitchell, Ind. Eng. Chem. 40 (4) (1948), 592

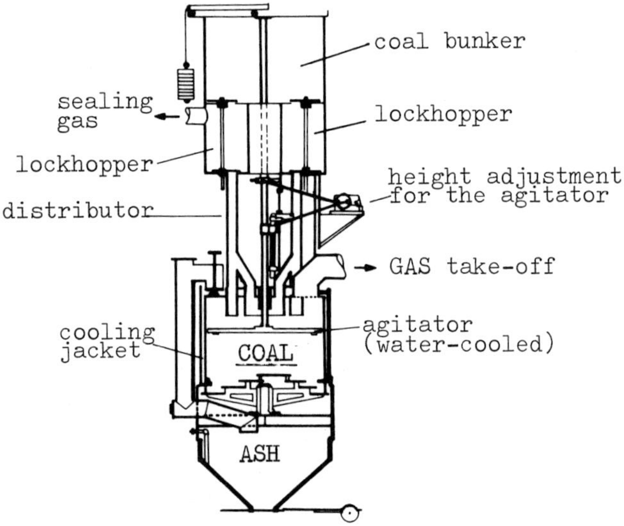

Figure 3 Wellmann-Galusha gas generator (with agitator)

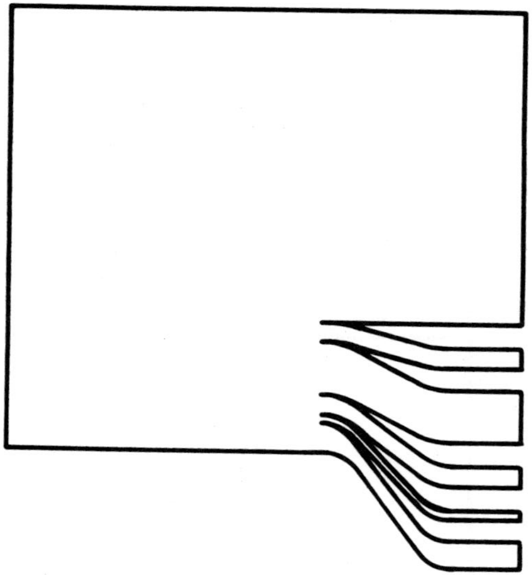

Figure 4 Heat balance for the Wellmann-Galusha process (with air)

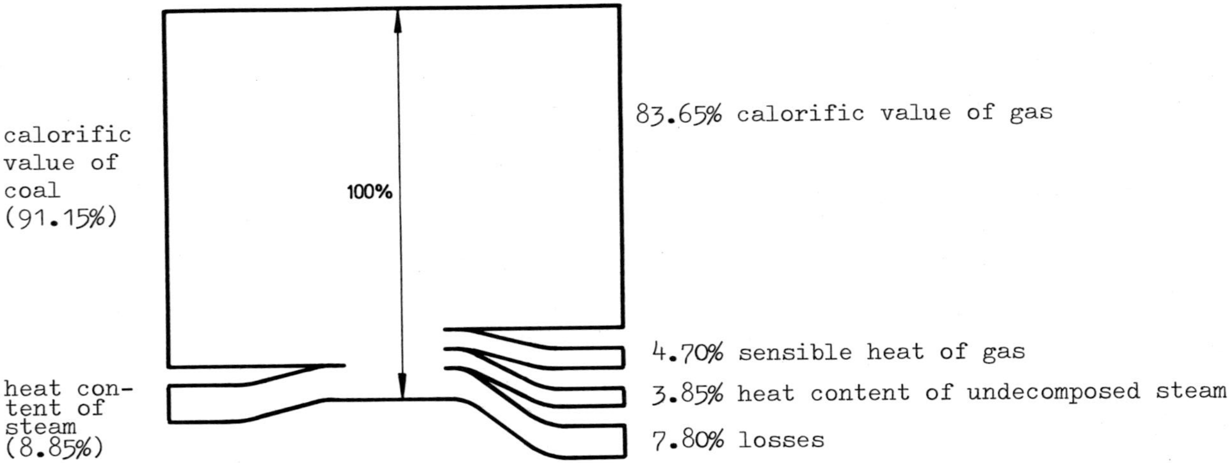

Figure 5 Heat balance for the Wellmann-Galusha process (with O_2/steam)

2. Lurgi pressure gasification process

I. Gasification system

Autothermic counter-current gasification under pressure of lump coal with steam and oxygen in a fixed-bed reactor.

II. Performance

Many Lurgi gasifiers have been used successfully on a large industrial scale in several commercial installations. This process, which was developed before the war, is the only pressure gasification process available which has so far been operated on a large industrial scale. Although many stages of the process had to be specially developed, all the details are now well established. For economic reasons the only installations in operation today in the West are the large plant at Sasolburg (Republic of South Africa), and a large experimental plant at Westfield (Scotland). The technical data given here are those published by Lurgi as the average values for such large plants, or operating results from Sasolburg. Operating figures for the Steinkohlengas AG plant at Dorsten, which was operated from 1955-1967, have also been included insofar as they are available.

Generator diameter:	about 3.6-4 m
Coal throughput:	15 t/h
Gas production:	max 55 000 m^3(iN)/h (dry)
Specific coal throughput:	1.2 t/m^2 h
Specific gas production:	max 4500 m^3(iN)/m^2 h

These are average values for recent operation at Sasolburg under optimized conditions. Lurgi has also designed reactors with a diameter of 4 m and a coal throughput of 17 t/h, corresponding to a raw gas production of about 55 000 m^3(iN)/h (dry), for large

SNG plants in the USA. Further development is aimed at generators with a raw gas production of 75 000 m^3(iN)/h.

Another area of further development is the operation of the generators with air instead of oxygen. At the Kellermann power station at Lünen tests are being run in a demonstration plant with an air-blown Lurgi generator integrated in a combined gas/steam turbine power plant; however, no detailed operating data for this mode of operation have yet been published.

It has recently been decided to bring the Lurgi plant at Dorsten (shut down in 1967) back into operation for experimental purposes. It is intended to develop the generator for use at higher pressures (Ruhr 100 plant). Other generators based on the Lurgi principle are in operation in Europe, in Yugoslavia, Czechoslovakia and the German Democratic Republic (Schwarze Pumpe Combine: about 30 generators).

III. Feed coal requirements

a. Rank

The Sasolburg plant uses a non-caking hard coal containing 19.7% moisture and 22.3% volatile constituents (dry basis).

Analysis (dry basis):	C	50.8%
	H	2.8%
	O	8.8%
	N	1.2%
	S	0.5%
	Ash	35.9%

At Westfield, and in the large-scale plants formerly operated at Dorsten and Coleshill, use was made of weakly caking coals with a swelling index (ASTM) of ≤ 2. For the large-scale US projects it is planned to use strongly caking coals, eg Pittsburgh coal with a swelling index (ASTM) of 6.5 and a caking index (DAMM) of 28. Gasification of similar coals has been tested successfully at Westfield: the gas-impermeable layer formed in the gasifier as a result of caking is continually broken up by stirring arms added as a modification. The Lurgi view is that there should be only very few types of coal, with extreme compositions, which cannot be used in the Lurgi pressure gasifier.

b. Particle size

It was found in Sasolburg that coal with a particle size of between 6 and 40 mm gives the best results. Tests in various plants have shown that ungraded coal can be used, although it requires some modification of normal operating conditions; gasifier output is also appreciably reduced. Experiments with Sasol coal showed that the proportion of fines (< 1 mm) should not exceed 7%.

c. Ash

The upper limit for the permissible ash content normally lies at about 35%, though coals containing more ash have been gasified. The coal should have a minimum ash content. The softening point of the ash (1340°C for Sasol coal) should be as high as possible: this governs the steam/oxygen ratio at which the reactor can be operated. The lower limit for the softening point of the ash is about 1000°C.

d. Coal pretreatment

For optimum operation the raw coal must be crushed, graded and, if necessary, dried. The upper limit for the moisture content of the coal is about 30%. The total dross content should not, however, exceed about 45%. Ungraded coal can be used, at the expense of output, provided that the proportion of fines is not too high.

IV. Raw gas composition (dry, tar-free basis)

	"typical"	Sasol	Dorsten
H_2	38 vol %	40 vol %	39.1 vol %
CO	24 "	18 "	22.8 "
CO_2	28 "	31 "	27.4 "
CH_4	10 "	9 "	9.6 "
N_2	not stated	1 "	0.5 "
H_2S	not stated	1 "	0.5 "
Calorific value:	11.8 MJ/m^3 (iN) (= 2830 kcal/ m^3(iN))	11.2 MJ/m^3 (iN) (= 2670 kcal/ m^3(iN))	11.7 MJ/m^3(iN) (= 2788 kcal/m^3 (iN))

At Sasolburg the gas is used mainly as synthesis gas, and also partly as town gas after CO_2 scrubbing. The by-products (tar, gas liquor and naphtha) are processed further to give marketable products.

V. Description

The raw coal is first washed, crushed, graded and (if necessary) dried. It is then fed into the gas producer through a periodically operating lockhopper and distributed mechanically across the cross-section of the shaft. For caking coals, the distributor is fitted with an additional attachment which loosens the coal bed while it is being preheated and devolatized, to prevent it from caking together into a large slab of coke.

Gasification is carried out at about 30 bar with a mixture of oxygen and steam, which is blown into the reactor through the revolving grate. The bed of ash resting on the grate serves to distribute the gasifying medium evenly and to preheat it. The gasifying medium is heated to the reaction temperature (according to Rudolph about 600-700°C) in a shallow combustion zone. This is followed by a zone in which mainly gasification by steam occurs, and in the next layer the coal is devolatized and gasified by hydrogenation (at 500-600°C). Finally, the raw gas dries the coal which is gravitating slowly downwards countercurrent, before leaving the gasifier at a temperature of between 300 and 800°C, depending on the type of coal. The gasifier consists of a double-walled vessel cooled by evaporation of water under pressure. The ash is periodically removed through a lockhopper at the base.

The raw gas leaving the reactor is quenched with water to 180-200°C in a scrubber-cooler, which washes out entrained coal dust and tar. It is then passed to a waste heat boiler in which low-pressure steam at 5-7 bar is obtained and the gas temperature falls to about 20-30°C. Part of the gas liquor condensed in this boiler is returned to the scrubber-cooler, in which a light oil condensate separates out. Surplus gas liquor and the suspension of tar and dust from the scrubber-cooler are worked up together in a tar separator, and the mixture of tar and dust is returned to the gasifier at the top of the coal bed. It acts as an effective dust catcher, and is subsequently cracked and gasified.

The composition of the raw gas is governed by the conditions under which the gasifier is operated. Since these conditions are chosen so as to optimize the gasifying process, the resulting H_2/CO ratio is usually unsuitable for the purposes of subsequent use. In collaboration with Steinkohlengas AG, Lurgi has therefore developed a conversion process which is applied to the raw gas before its purification proper, which is thereby simplified. The catalyst used for this also promotes hydrogenation, so that any higher hydrocarbons or hydrogen cyanide remaining in the raw gas are hydrogenated; in particular, gas naphtha is desulphurized, which makes subsequent treatment considerably easier. It is not necessary to add extra steam to bring about conversion, since the gasification steam which remains undecomposed is sufficient to reduce the CO content to 4.5%. The reaction is carried out in two stages at 440°C and 400°C in a bypass stream of the raw gas. The waste heat is used to produce low-pressure steam. It is a particular advantage that no additional equipment is needed to convert the raw gas.

Further purification of the gas is effected by a Rectisol wash, usually in two stages. The scrubbing liquids are organic solvents, preferentially methanol, at temperatures between 0°C and -50°C (-55°C at Sasolburg). The first stage removes gas naphtha and other higher hydrocarbons as well as residual water vapour and ammonia. The solvent is stripped of dissolved organic substances and recovered by distillation. In the second stage H_2S, COS and CO_2 are removed, the solvent being the high-purity methanol from the distillation. It is stripped of the acid gases by hot regeneration and then recycled for use in the first stage without further treatment.

For sulphur recovery in a Claus kiln, it is often also necessary (depending on the sulphur content of the raw coal) to fractionate the acid gases so as to increase the content of hydrogen sulphide.

At Sasolburg, continuous industrial production yields a synthesis gas with a residual H_2S content of less than 0.2 ppm (corresponding to 0.3 mg H_2S/m^3(iN)).

The dynamic behaviour of the gasifier has not been reported. It was found at Sasolburg that throughput could be increased by about 30% in order to meet transitory peak requirements: no lower limit for the load could be found in the literature.

It has also been reported from Sasolburg that the start-up time for a reactor (up to its rated capacity) is now about 12 h, compared with what was previously 3 days, and that the average availability of the gasifier is now 83%.

VI. Production of gasifying medium

Oxygen: Fractionation of air by the Linde-Fränkl process

Steam: Part (10-15%) of the steam for gasification is obtained from the pressurized cooling-water jacket of the generator. The low-pressure steam from the waste heat boilers following the scrubber-cooler and the gas converter cannot be used for gasification. The process is not self-sufficient in steam.

VII. Technical data

	"typical" (A)	Sasol (B)	Dorsten (C)	unit
a. Carbon conversion efficiency:	not stated	not stated	99.6	%
b. Gasification efficiency:	85	80.8	77.7	%
c. Thermal efficiency:	68-70	75.5	not stated	%
d. % steam decomposition:	ca 36	not stated	ca 40	%

	"typical" (A)	Sasol (B)	Dorsten (C)	unit
e. Inputs				
Oxygen:	206	216	299	m^3(iN)/t coal
	0.15	0.16	0.175	m^3(iN)/m^3 (iN)raw gas
	13.0	14.3	14.7	m^3(iN)/GJ raw gas
Steam:	1.20	1.0	1.38	t/t coal
	0.75	1.07	0.805	kg/m^3(iN) raw gas
	63.1	66.6	67.4	kg/GJ raw gas
Coal:	0.63	0.74	0.585	kg/m^3(iN) raw gas
	52.5	66.4	49.0	kg/GJ raw gas
f. Steam/oxygen ratio:	5.8	4.7	4.6	kg/m^3(iN)
g. Products				
Gas yield	1595	1350	1708	m^3(iN)raw gas/t coal
Tars, oils, naphtha, etc:	(90)	23.6	73.4	kg/t coal
Crude phenols:		(3.8)	9.1	kg/t coal
Fertilizer (calculated as N):	not stated	8.5	7.3	kg/t coal
Low-pressure steam:	not stated	2.0	not stated	GJ/t coal
h. Gasification pressure:	(30)	not stated	23.3	bar
Gasification temperature:	760-870	not stated	800-900	°C
Exit gas temperature:	370-600	not stated	550	°C

	"typical" (A)	Sasol (B)	Dorsten (C)	unit
State of gasifying medium:				
Oxygen:	32	not stated	27.4	bar
Steam:	35.5	"	27.4	bar
	400	"	500	^oC
i. Water consumption:	not stated	not stated	not stated	
j. Other energy consumption:	4.69	5.86	not stated	GJ/t coal
k. Gas purification and treatment of by-products:	-	-	-	

The purification of the gas has already been outlined in Section V. The raw gas is first quenched in a scrubber-cooler, in which higher-boiling tar fractions are condensed together with suspended dust and ash. The gas is then passed to a waste heat boiler, where gas liquor containing light oil and phenol is condensed: some of this is transferred to the scrubber-cooler, and the surplus is routed directly (together with the tar suspension from the scrubber-cooler) to a separator. The tar thus obtained is returned to the gasifier for dust filtering and for gasification: alternatively, it can be removed if other uses are intended. The gas liquor is treated in the Phenosolvan plant.

A bypass stream of the raw gas is converted in order to obtain the desired H_2/CO ratio, and is then reunited with the remainder and cooled. An oil-water suspension is deposited, which is separated in a settling tank. The oil can either be used for heating or can be processed to give motor fuel; the water layer is added to the gas liquor previously obtained and routed to the Phenosolvan plant.

As already described, further purification of the gas employs a Rectisol wash, giving as by-products more gas naphtha and a mixture of acid gases. The acid gas is converted to sulphur in a Claus kiln and then, where required, to sulphuric acid.

The Phenosolvan plant is used for the treatment and purification of effluent which contains phenols. The gas liquor from gas purification contains the water-soluble impurities from the raw gas, mainly crude phenols, ammonia and tar acids. It is stripped of suspended matter by filtration and then extracted several times in a counter-current extractor with isopropyl ether (butyl acetate is used instead at Sasolburg). The extract is separated in a distillation column and the solvent is recycled. The liquor is now free from organic impurities but still contains ammonia and solvent. After preheating, these are removed by cyclic stripping in a desorption column. The resulting purified effluent now has a residual content of 20 ppm of phenol (250 ppm at Sasolburg) and 60 ppm of ammonia (240 ppm at Sasolburg).

The ammonia obtained as a by-product can be combined with sulphuric acid to give ammonium sulphate fertilizer.

VIII. Construction and development

LURGI Mineralöltechnik GmbH, Frankfurt/Main

Operator

South African Coal, Oil and Gas Corporation Ltd, Sasolburg, Republic of South Africa

References

(1) P. Rudolph, 4th Synthetic Pipeline Gas Symposium, Chicago, Illinois, 30-31 October 1972

(2) H. Hiller, DGMK Conference, Hamburg, 1 October 1974

(3) J.C. Hoogendoorn, Clean Fuels from Coal Symposium, Chicago, Illinois, 10-14 September 1973

(4) LURGI Handbook, 2nd edition, Frankfurt 1972

(5) ibid; Gasette, house journal of Ruhrgas AG, March 1974, p 6

(6) P. Rudolph, FAI Symposium, Coal as Feedstock for Fertilizer Production, New Delhi, India, 29-30 April 1974

(7) G. Baron, Chem. Ing. Techn. 46 (1974), 943

(8) F. Bieger, paper presented at Symposium on the Pressure Gasification of Coal, Portorož, October 1964

(9) J.A. Gray, Coal and Energy Quarterly No 11 (1976), 15/22

(10) F.J. Dent, BCURA Quarterly Gazette No 42 (1960), 1

(11) H. Puhr-Westerheide, VGB Kraftwerkstechnik 54 (1974), 532

(12) J.C. Hoogendoorn, Communication 960, presented at the 112th Annual Meeting London, May 1975

(13) 13th World Gas Conference London 1976, Report of the Committee on Production of manufactured Gases, IGU/B-76

(14) G. Röbke, Conference on Kohlevergasung und Hydrierung (Coal Gasification and Hydrogenation), Haus der Technik, Essen, 7-8 December 1977

(15) C. Hafke, Techn. Mitt. (HDT) 71 (1978), p 142

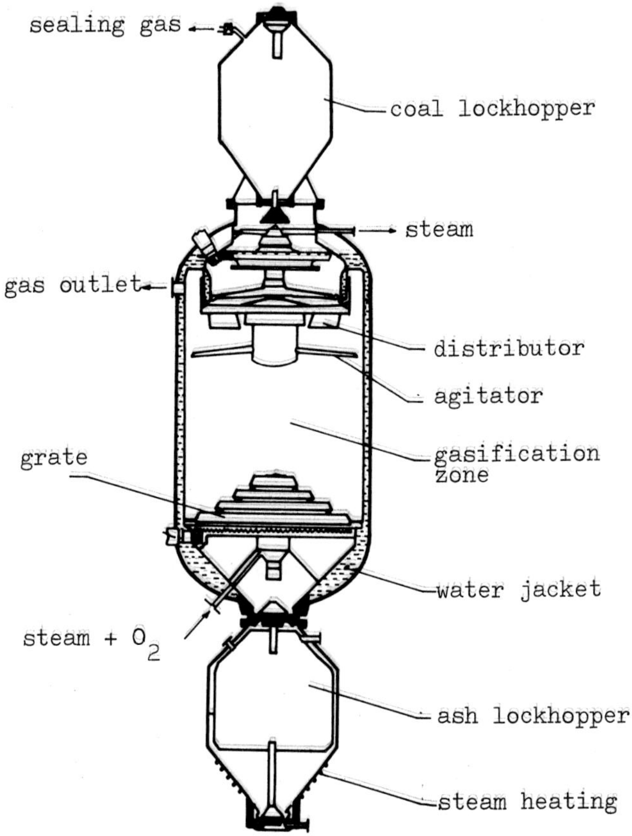

Figure 6 Lurgi generator for coal gasification, fixed-bed, 25 bar

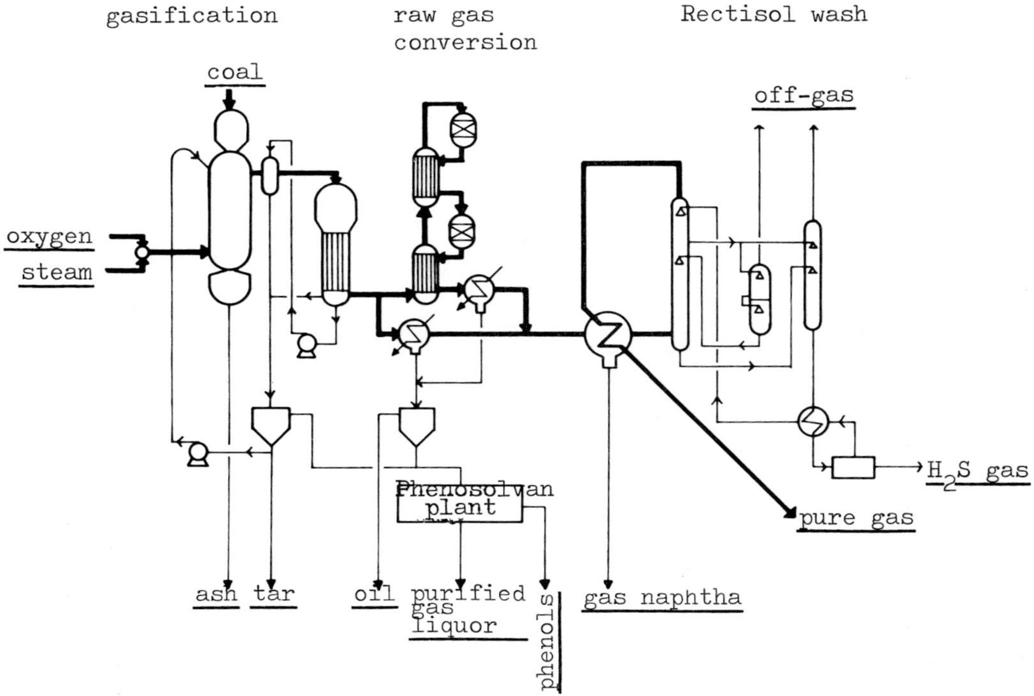

Figure 7 Lurgi process for pressure gasification of coal (Lurgi-Mineralöltechnik GmbH)

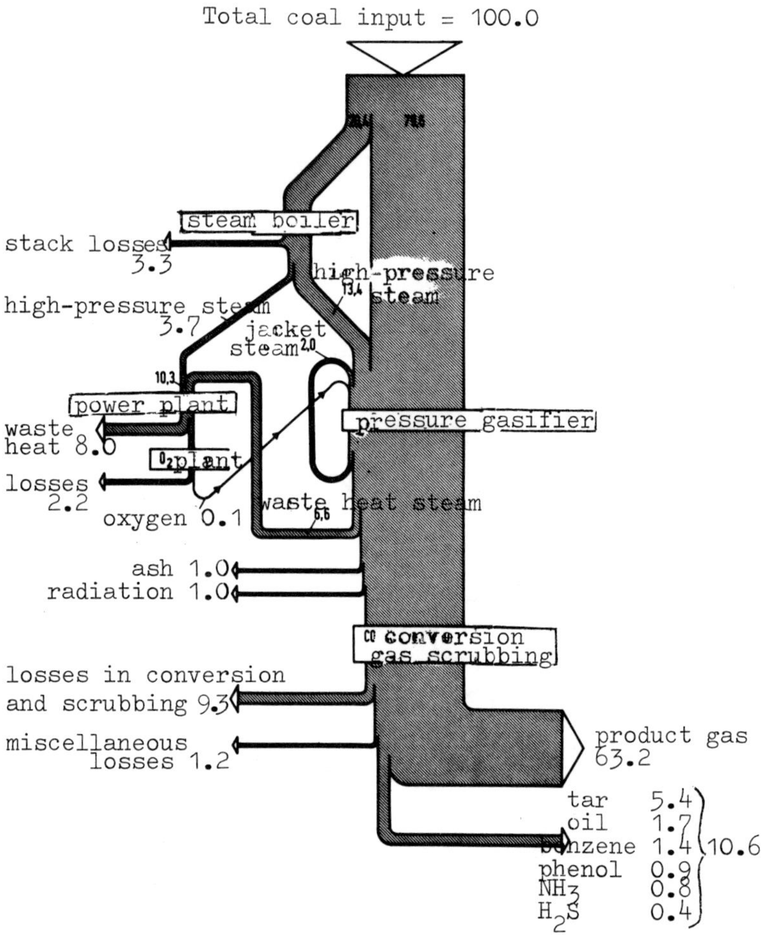

<u>Figure 8</u> Energy flow diagram for Lurgi pressure gasification (F.J. Dent)

3. Winkler process

I. Gasification system

Autothermic gasification in a fluidized bed with steam and air or oxygen at atmospheric pressure.

II. Performance

In all, 16 large-scale industrial plants have been built with 36 gasifiers having rated capacities of between 5 and 60 x 10^3 m^3 (iN) of synthesis gas per hour, corresponding to a coal throughput of about 3-35 t/h. At present a number of large-scale plants are still in operation.

The data given here are mean values from the operating figures of large-scale plants.

Cross-section of shaft:	between 6.6 and 25 m^2
Height of shaft:	up to 21.5 m
Specific output:	1100-3000 m^3 (iN)/m^2h

III. Feed coal requirements

a. Rank: The process is suitable chiefly for lignites, since they are highly reactive. Reactive hard coals have also been gasified, as well as semi-coke and coke breeze, though these require a higher temperature (less favourable heat balance). Moderately caking coals can also be gasified. Coals containing between 4 and 46% ash and between 6 and 47% volatile constituents have been used successfully.

b.	Particle size:	May be up to about 8 mm. The coal must be crushed but need not be graded. A large dust fraction (eg 60% between 0 and 1 mm) does not cause any difficulties, but leads to considerable losses in the form of entrained dust.
c.	Ash:	A high or very high ash content does not interfere directly, but impairs the heat balance. The softening point must be higher than the reaction temperature (≈ 800°C for lignite), and should be above 1200°C preferably. The ash content may vary during operation.
d.	Coal pretreatment:	Crushing without grading. Drying, if necessary, sufficient to ensure satisfactory mechanical transport (18% moisture content for lignite). At one time drying to 8% moisture was common.

IV. Raw gas composition and calorific value (dry basis)

There are considerable variations depending on the reaction conditions chosen and the type of coal used. The following composition can be taken as an average:

CO_2	13 -25 %
CO	30 -50 %
H_2	35 -46 %
CH_4	1 - 1.8%
N_2	0.5- 1.5%

H_o = 9.2-13.0 MJ/m^3(iN) (≙ 2200-3100 kcal/m^3(iN))

Using air for gasification, the Leuna works has produced a power gas with a calorific value of

4.58 MJ/m^3(iN) (≙ 1095 $kcal/m^3$(iN)). Since Winkler gas always contains a relatively high proportion of $CO + H_2$, almost all plants have been built to supply synthesis gas to fertilizer factories and Fischer-Tropsch synthesis works. As far as is known, power gas is produced only at the Leuna works, air being used instead of oxygen. It is also planned to use the gas in direct reduction plants.

V. Description

Since the Winkler process has a wide tolerance on incoming fuel specification, pretreatment of the coal is very simple and straightforward. Even very fine material can be processed, so that simple crushing without grading is usually sufficient. The carbon conversion efficiency does, however, depend on suitable particle size. If the particle size used is too fine, the consumption of fuel in losses of entrained dust can be as high as 50% or more when the generator is heavily loaded. Apart from the large proportion of entrained and incompletely gasified dust, this causes considerable overloading of the dust extractors and increases corrosion. A fuel with too high a proportion of large particles also causes increased losses, since the mixing which occurs in the fluidized bed means that large particles are removed with the ash at too early a stage, ie still with a high content of carbon.

After crushing the coal, partial drying may be necessary in order to prevent agglomeration in the conveying machinery or the bunkers.

The fuel in the bunkers is kept under a slight excess pressure of protective gas (CO_2) in order to prevent flashback. From here it is fed continuously into the generator via screw conveyors. The generator consists of a brick-lined shaft with an iron sheath. Its lower part is a truncated cone, sometimes incorporating a grate chamber through which the primary gasifying medium is introduced. In more recent generators the grate is usually omitted and the gasifying medium is introduced through water-cooled jets.

The gas produced in the fluidized bed is first mixed with the secondary gasifying medium in a ring main 3-4 m above the coal bed, thus gasifying some of the suspended dust carried out with it. A waste heat boiler built into the shaft ensures that the softening temperature of the ash is not exceeded. Before use, the gas then passes successively through another waste heat boiler, a feed-water preheater, a multiclone, a scrubber-cooler and a drip catcher.

If the coal dust collected in the waste heat boilers and the cyclones cannot be conveniently used for firing, it is returned to the fluidized bed. The ash is transported by discharge screw conveyors into storage bins, which are periodically emptied. Because of the intensive mixing in the fluidized bed the reactive fuel and the unreactive ash are not spatially separated in the Winkler generator. The ash removed therefore represents part of the incandescent bed, and has roughly the same reactivity. Furthermore, the C-content of the fluidized bed must not fall below about 5%, in order to prevent any dangerous break-through of oxygen. This means that the ash contains 5% ungasified material, and a continuous and careful watch must be kept on its C-content.

At one time, the generator was ignited by means of a separate firing generator, which was first fired with wood in order to produce an incandescent bed of coal.

The incandescent coal was then transferred to the main generator through a connecting pipe, after which the fluidized bed was set up. In more recent plants ignition takes place in the main generator itself, either by initial firing with wood, as previously, or by preliminary heating with gas burners.

The generator can be shut down extremely simply and rapidly. After the oxygen is turned off the supply of steam can be reduced. At the same time the system is flushed out with a protective gas (CO_2) as far as the cyclones, in order to prevent access of air to parts of the system which may still contain reactive dust. Steam can then be injected again, plus air, in order to burn the generator empty, the gas being flared off. If the generator is to be kept hot, CO_2 is blown in instead of steam and air so as to cool the feed screw conveyors; the bed remains incandescent for some time. If the generator is to be put on stand-by for a longer period the bed must be re-heated from time to time by air blowing.

The Winkler generator is extraordinarily flexible in adapting to the amount of product required at any given moment. In a very short time it can be switched from minimum load ($\approx$ 25% of rated load) to maximum load ($\approx$ 150% of the rated load); in this respect it is superior to all other known processes for gasifying coal. These changes of load cause relatively little decline in efficiency, but an increasing fines fraction makes the carbon conversion efficiency more dependent on the load, because of the greater losses of suspended dust.

A new development envisages the conventional gasification of dried lignite with oxygen or preheated air by the Winkler process, but at pressures between 1.5 and 5 bar. Experiments have shown that adding limestone to the fluidized bed prevents the ash from caking together by raising its softening or melting point. This <u>High-Temperature Winkler</u> (HTW) process can thus be operated with a reaction temperature which is higher than that previously attained in the industrial gasification of lignite on a large scale by the Winkler process. The idea behind this development is to use the higher operating temperature to produce a gas with a high reducing potential.

The hot raw gas can then be passed directly into an adjacent direct-reduction apparatus for making iron sponge. A suitable gas for this purpose can be produced by gasifying predried lignite at about 1000°C, where the higher gasification temperature increases the reducing potential by favouring the production of carbon monoxide and hydrogen. The addition of limestone not only raises the softening point of the ash but also desulphurizes the gas in the fluidized bed itself. As a result of the addition of limestone, the ash is suitable for making cement.

VI. Production of gasifying medium

Oxygen: Conventional air fractionation plant

Steam: The steam for gasification is taken almost entirely from outside sources, while the high-pressure steam from the waste heat boilers is used elsewhere. This high-pressure steam is probably sufficient to run the air fractionation plant.

VII. Technical data

a. Carbon conversion efficiency: 55-90%

b. Gasification efficiency: 45-75%

c. Thermal efficiency: up to 82%

d. % steam decomposition: 35-45%

e. Inputs:

Oxygen: 0.52 t/t feed coal = 365 m^3(iN)/t coal

0.23 m^3(iN)/m^3(iN) synthesis gas

20.8 m^3/GJ in the synthesis gas

Steam: 0.88 t/t feed coal
0.55 kg/m^3 (iN) synthesis gas
0.045 t/GJ in the synthesis gas

Coal: 0.64 kg/m^3 (iN) synthesis gas
0.057 t/GJ in the synthesis gas

f. Steam/oxygen ratio: 2.4 kg/m^3 (iN)

g. Products

Gas yield: 1580 m^3(iN) synthesis gas/t coal

Steam produced: 0.045 t/GJ in the synthesis gas = 6.0 t/h at an output of 12 000 m^3/h

h. Gasification pressure: Atmospheric; recent experiments envisage operation under pressure

Gasification temperature: 850-1100°C

Exit gas temperature: 800-1000°C

State of gasifying medium: oxygen 1.4 bar
steam 3-4 bar, 350-500°C

i. Water consumption: Estimated water consumption for steam production is 48 kg/GJ = 6.3 t/h at an output of 12 000 m^3/h. Recycled condensate (used together with low-pressure steam for regasification): 9.9 kg/GJ = 1.308 t/h at 12 000 m^3/h

Water consumption for scrubbing: not stated

j. Electricity consumption: 6.64 kWh/GJ = 878 kW at an output of 12 000 m^3/h

k. Gas purification: In the Winkler process the purification of the gas is complicated by the very high dust content of the raw gas. In most existing plants initial dust separation takes place in the waste heat boiler, and this is followed by multiclones. The separated dust, which still contains 40-60% combustible material, can either be used for coal-dust firing, or be returned to the generator. Some of the dust is also used as an adsorbent on account of its high surface activity.

After dry dust removal, the gas still has a suspended dust content of 10-40 g/m^3. Most of this is removed in the scrubber-cooler which follows (residual content about 1 g/m^3). Final dust removal is often carried out by disintegrators, which reduce the dust content to less than 10 mg/m^3. Alternatively, wet electro-filters are used, which require far less expenditure on electricity, or venturi scrubbers, which have a considerably lower water consumption. Conventional methods are used for desulphurizing the gas, depending on external conditions and the quality of gas required.

VIII. Construction and development

Rheinische Braunkohlen AG, Cologne (HT-Winkler)

Davy Power Gas GmbH, Cologne

Operators

A. Fabrika Azotnih Jendinjenjy, Gorazde, Yugoslavia

B. Azot Sanyyii TAS, Kütahya, Turkey

C. Neyveli Lignite Corporation, Madras, India

D. several plants in the German Democratic Republic

References

(1) W. Flesch, Glückauf 90 (1954), 537

(2) W. Flesch and G. Villing, Erdöl und Kohle 15 (1962), 710

(3) I.N. Banchik,"Clean Fuels from Coal" Symposium, Chicago, 10-14 September 1973

(4) J. Wilsdorf and H.G. Bürgel, Chem. Techn. 26 (1974), 329

(5) Anon, booklet published by Davy Power Gas/Bamag Chemietechnik Chemiebau Zieren 1976

(6) G. Anwer and F. Bögner, Brennstoff Wärme Kraft 28 (1976), p 57

(7) J. Anwer, F. Bögner and E. Wittstock, Brennstoff Wärme Kraft 28 (1976), p 189

(8) I.N. Banchik, T.K. Subramaniam and J.K. Marteen, Hydrocarbon Processing, p 6 (1977) No 3 (March) 121

HT Winkler process

(1) M. Meraikib and F.H. Franke, Chem. Ing. Techn. 42 (1970), p 834

(2) F.H. Franke, W. Wenzel, B. Dreyer and W. Sitte Braunkohle 26 (1974), p 178

(3) F.H. Franke, Conference on "Kohlevergasung und -hydrierung" (Gasification and hydrogenation of coal), Haus der Technik, Essen, 7-8 December 1977

Figure 9 Winkler process (BAMAG Verfahrenstechnik GmbH (Davy Power Gas, Inc.))

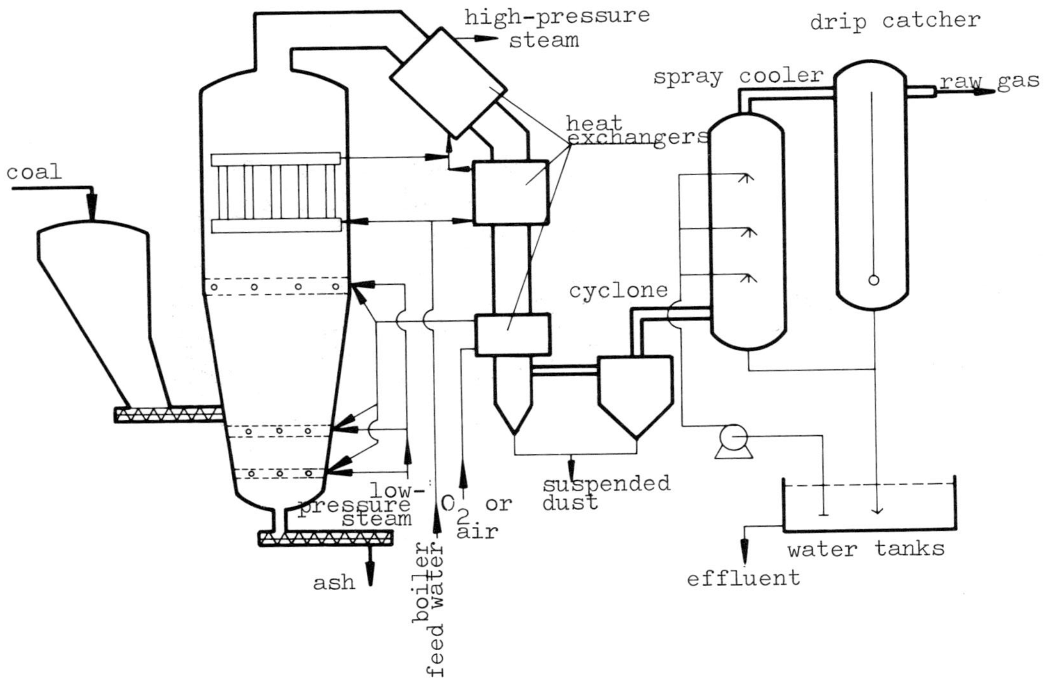

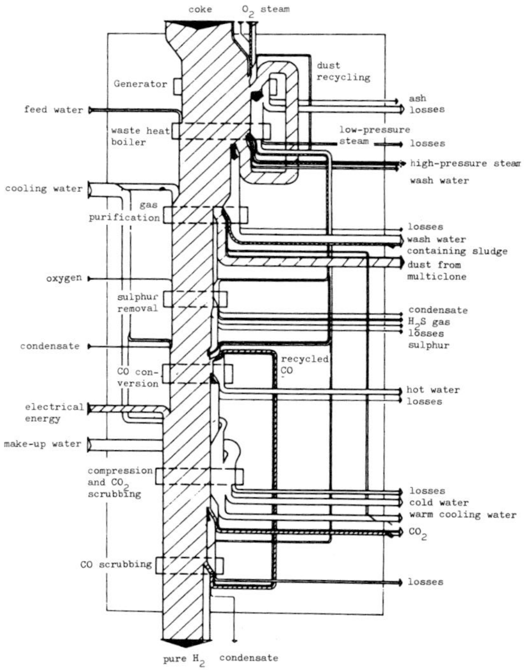

Product flow	Energy in %	Exergy in %
Overall balance		
Inputs		
Carbonization coke	80.2	86.7
Oxygen	0.1	0.0
Steam (4.5 atm)	7.3	2.1
Feed water	2.5	0.3
Electrical energy	9.9	10.9
Total	100.0	100.0
Useful products		
Pure hydrogen	54.4	50.5
Condensate	0.1	0.0
Steam (22 atm)	3.3	1.3
Multiclone dust	11.0	12.3
H_2S gas	1.0	0.4
Sulphur	0.1	0.1
Total	69.9	64.6
Losses	30.1	35.4
Breakdown of losses		
External losses		
Dust in wash water containing sludge	5.0	4.7
Heat in cooling water	17.5	1.4
Wash water	3.7	0.8
Ash	0.9	0.8
CO_2	13.3	11.3
Total	40.4	19.0
Internal losses		
Generator	25.8	39.4
Waste heat boiler	7.3	16.2
Gas purification	2.3	5.5
Sulphur removal	2.7	1.7
CO conversion	1.4	12.1
Compression	13.7	4.2
CO scrubbing	6.4	1.9
Total	59.6	81.0
Exergetic index for inputs		0.904
Exergetic index for useful products		0.835

Figure 10 Energy flow-sheet for hydrogen production by the Winkler process

4. Koppers-Totzek process

I. Gasification system

Autothermic entrained gasification by oxygen and steam at atmospheric pressure flowing co-current. Further development to use higher gasification pressures is in progress.

II. Performance

So far, 53 gasifiers have been built for 14 large installations, but 14 of these operate with oil or gas as fuel. Gasifiers with 2 burner nozzles predominate at present. They are about 7.5 m high and have an internal diameter of up to 3.5 m. Their internal volume is about 28 m^3. Gasifiers with 4 burner nozzles have about twice this volume.

Some of the following numerical values are mean values from the operating figures of large-scale installations. The technical data in Section VII come mostly from the designs published by Koppers-Pittsburgh for large installations.

Coal throughput:	13.5 t/h (with 2 burner nozzles)
	20.8 t/h (with 4 burner nozzles)
	34.4 t/h (with 4 burner nozzles - planned)
Raw gas production:	21 000 m^3(iN)/h (with 2 burner nozzles)
	32 300 m^3(iN)/h (with 4 burner nozzles)
	55 000 m^3(iN)/h (with 4 burner nozzles - planned)

An experimental plant for gasification under pressure, in operation since the end of 1976, has a throughput of 250 kg coal per hour. A pilot plant (pressure gasification of 6.25 t coal/h) has come into operation in the meantime.

III. Feed coal requirements

a. Rank:	Any coal can be used, and also petroleum coke or oil.
b. Particle size:	70% $<$ 0.075 mm, or not more than 10% residue when passed through a 0.09 DIN screen.
c. Ash:	Upper limit about 40% of the dried feed coal. (The softening point of the ash is not critical, since it is melted completely.) A slagging flux may need to be added.
d. Coal pretreatment:	Drying and pulverizing, moisture content 2-8%

IV. Raw gas composition and calorific value (dry basis)

CO	55 -66 %
CO_2	7 -12 %
H_2	21 -32 %
N_2	1 %
CH_4	0.1%
H_2S	0.5- 1 %

H_o = 10.6-11.8 MJ/m^3(iN) = 2540-2820 $kcal/m^3$(iN)

With very unreactive fuels (eg coke) the CO content is very high, and the H_2 content therefore low. With reactive fuels the reverse is the case (the extreme case is natural gas: 34% CO, 61.3% H_2).

V. Description

Although the Koppers-Totzek process is largely independent of the rank of coal, the pretreatment of the fuel must be adapted to each type of coal. The coal is pulverized to a particle size of less than 75 μm, and at the same time is dried with flue gas from the combustion of product gas to a residual moisture content of 2-8%, depending on the type of feed coal. It is then conveyed pneumatically into the coal bunker by means of the nitrogen obtained in the air fractionation plant, and fed periodically into the charging hoppers of the gasifier.

The coal dust is transported to a mixer head by screw conveyors, which provide accurate volume control; from there it is taken up by a current of oxygen and steam and blown into the burner nozzles of the gasifier. Any flashback of flames through the conveying tubes between the mixer head and the burners is effectively prevented by the relatively low temperatures and high transport speeds, which exceed the rate of flame propagation for mixtures of coal dust and oxygen.

The high gasification temperatures (1500-1900°C) mean that no tarry or phenolic by-products are formed, which simplifies the purification of the gas considerably. About half of the ash runs down the walls of the gasifier as liquid slag; it is quenched in a tank of water and removed from beneath the gasifier in granular form. The other half is entrained as suspended ash in the gas produced. To prevent caking up in the waste heat boiler, as it leaves the gasifier the gas is cooled to below the melting point of the ash by water sprays.

The product gas first passes through a waste heat boiler, which produces high-pressure steam (up to 100 bar, depending on the design); this is followed by dust removal in two venturi scrubbers. The amount of steam produced is more than sufficient to run the compressors of the oxygen plant. The dust content of the gas leaving the venturi scrubbers is still about 5-15 mg/m^3 (iN). If there are stringent requirements for a low dust content in the gas (compressor quality) an electro-filter has to be incorporated, which reduces the dust content to 0.25 mg/m^3 (iN). The gas is then cooled and finally desulphurized by any conventional method, depending on the circumstances.

The gasifier output can be easily and rapidly varied in the range 60-100% by changing the feed speed of the screw conveyors. The plant can be shut down instantaneously, if desired. Since the gasifier's fuel charge is very small, a very rapid response is obtained to all adjustments made. The resumption of full gas production in a generator which has been shut down but kept warm takes about 30 minutes, though it takes somewhat longer to raise steam production to its full value.

Since the only moving parts of the gasifier are the screw conveyors, operation involves little wear or maintenance. The availability of the whole plant is usually stated to be better than 95%.

VI. Production of gasifying medium

Oxygen: A conventional air fractionation plant of sufficient size can be run with the high-pressure steam from the waste heat boiler. The nitrogen thus produced serves as a medium for conveying the feed coal, as a flushing gas when operation is interrupted, etc.

Steam: Overall, rather more low-pressure steam is produced in the jacket of the gasifier than is needed for gasification.

VII. Technical data

a. Carbon conversion efficiency: 90-96%, under pressure up to 98%

b. Gasification efficiency: 70-77%

c. Thermal efficiency: 75-85%

d. % steam decomposition: 85-90%

e. Inputs

Oxygen:	535	m^3(iN)/t feed coal (dry)
	0.325	m^3(iN)/m^3(iN) raw gas (dry)
	30.1	m^3(iN)/GJ raw gas
Steam:	236	kg/t feed coal (dry)
	0.143	kg/m^3(iN) raw gas (dry)
	13.3	kg/GJ in the raw gas
Feed coal (dried):	0.61	kg/m^3(iN) raw gas (dry)
	56.6	kg/GJ in the raw gas

f. Steam/oxygen ratio: 0.44 kg/m^3(iN)

g. Products:

Gas yield: 1640 m^3(iN) dry raw gas/t dried feed coal

Steam production:

Gasifier jacket: 11.2 kg/GJ in the raw gas (142°C, 3.7 bar)

Waste heat boiler: 67.6 kg/GJ in the raw gas (400°C, 45 bar)

h. Gasification pressure: atmospheric

Gasification temperature: 1500-1900°C

State of gasifying medium:

Oxygen: 98%, 100°C
Steam: 3-4 bar, 140°C

i. Water consumption:

Gasifier jacket: 11.7 l/GJ in the raw gas (110°C)
Spray cooling of gas: 8.4 l/GJ in the raw gas (32°C)
Waste heat boiler: 71.2 l/GJ in the raw gas (110°C)
Ash cooling: 71.7 l/GJ in the raw gas (from cooling tower)

j. Coal consumption for energy production

Losses in drying and pulverizing: 2.36 kg/GJ in the raw gas

The remaining energy requirement is largely covered by the steam produced in the plant; a further 10.7 kWh/Gcal of electricity are required.

k. Gas purification: As stated in Section V, the purification of the gas depends on the quality of gas required. The dust content can be reduced to about 0.25 mg/m^3(iN), and the sulphur content can be brought down to any desired value by modifying conventional procedures (multi-stage if necessary). It is unquestionably a special advantage of this process that no gases or liquors containing tar or phenols are produced, which very much simplifies both gas and water purification.

VIII. Construction and development

Krupp-Koppers GmbH, Essen, Federal Republic of Germany

Pressure gasification (Shell-Koppers process):

Shell Internationale Petroleum, Maatschappij BV, The Hague

Deutsche Shell AG, Hamburg

Krupp-Koppers GmbH, Essen

References

(1) F. Totzek and K.H. Osthaus, Brennstof-Chem. 31 (1950), 173

(2) K.H. Osthaus and T.W. Austen, Gas World, 12 January 1963, p 98

(3) K.H. Osthaus, VGB-Kraftwerkstechn. 54 (1974), 221

(4) J.F. Farnsworth, H.F. Leonhard, D.M. Mitsak and R. Wintrell, Paper presented at the 1974 Annual Convention of the Association of Iron and Steel Engineers, 22-24 April 1974, Philadelphia, Pa

(5) J.E. Franzen and E.K. Goeke, 6th Synthetic Pipeline Gas Symposium, 28-30 October 1974, Chicago, Illinois

(6) J.E. Franzen and E.K. Goeke, Hydrocarbon processing 55 (1976), p 134

(7) H. Staege, Glückauf 113 (1977), p 416

(8) J.E. Franzen, Techn. Mitt. Krupp. Forsch. Ber. 35 (1977), p 83

Pressure gasification

H.J. Kraaijveld, Conference on "Kohlevergasung und -hydrierung" (Gasification and hydrogenation of coal), Haus der Technik, Essen, 7-8 December 1977

M.J. van der Burgt and H.J. Kraaijveld, 175th ACS meeting, Anaheim, Calif. Industrial and Engineering Division, 16 March 1978

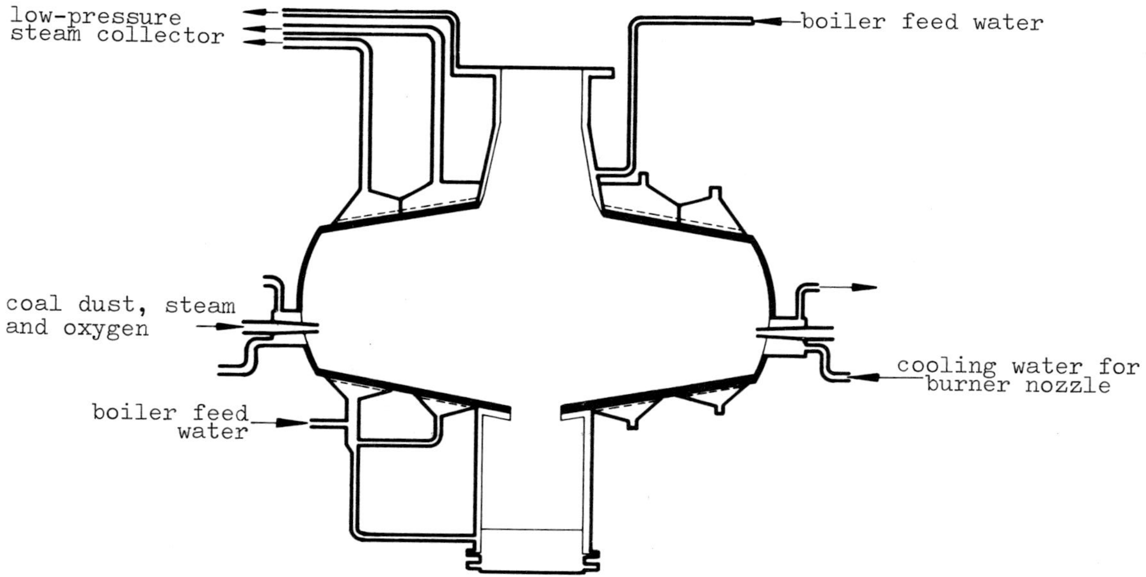

Figure 11 Koppers-Totzek gasifier: model with two burner nozzles

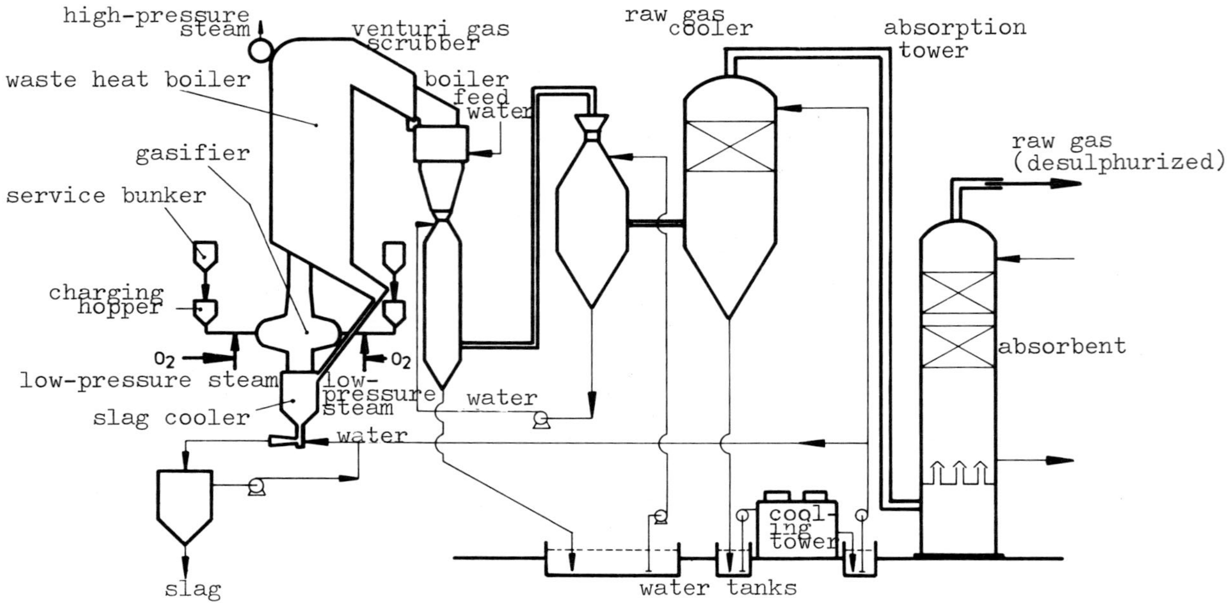

Figure 12 Koppers-Totzek process (Krupp-Koppers GmbH)

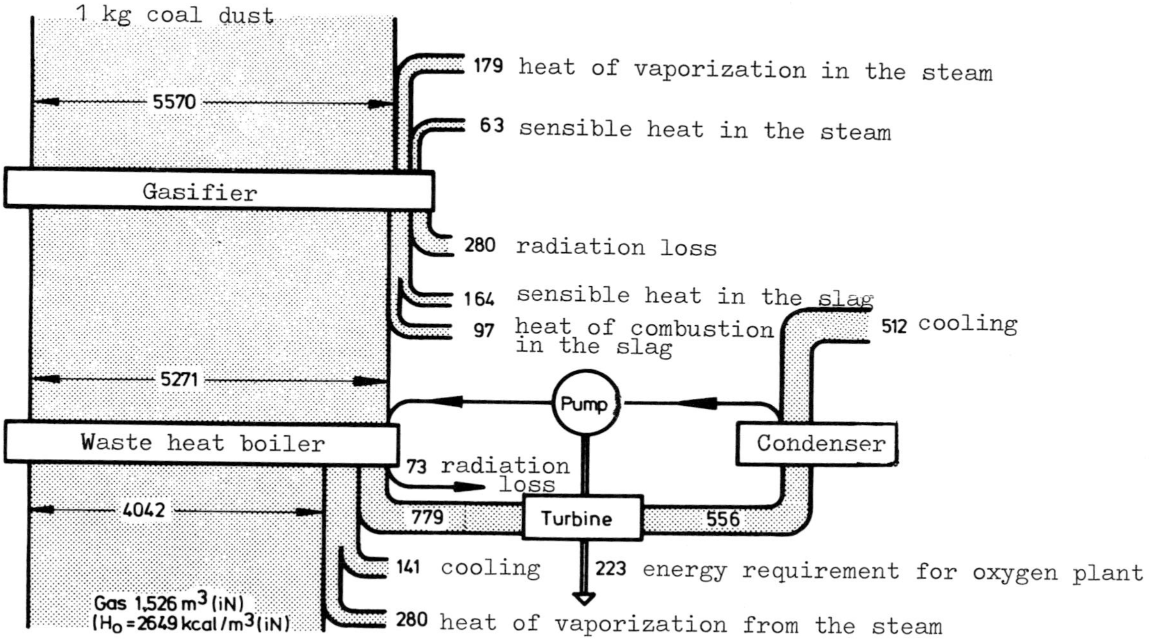

Figure 13 Heat balance for Koppers-Totzek gasification (figures given in kcal)

B. Processes not currently in commercial operation and processes under development

1. Hot-blast gas processes

In the early days of industrial coal gasification only coke and non-caking hard coals were used as raw materials. However, in the 1920s the demand for large quantities of synthesis gas and the high cost of using coke as a raw material directed attention towards the gasification of lignites. But before these soft coals could be used for gasification, suitable continuous processes had to be developed which could overcome the special difficulties associated with them, such as the low melting point of the ash. On the other hand, the high reactivity of lignites made gasification at low temperatures possible for the first time.

In this situation the development of the hot-blast gas processes proved to be a successful measure, since with lignites the sensible heat of the blast gas is sufficient to maintain the gasification reaction, while the danger of forming slag and clinker by overheating can be kept firmly in check. Furthermore, the process yields a continuous stream of a water gas which is largely free from nitrogen without using expensively-produced pure oxygen.

In hot-blast gas generators the coal is gasified with steam. The heat required to heat up the fuel and to supply the endothermic heat of reaction is introduced into the generator by a current of flushing gas or blast gas, which also contains the steam for gasification. Processes differ in the way in which the heat is actually supplied: heat is transferred to the blast gas when it passes through regenerators which precede the gasifier, and the regenerators themselves are heated by some external means, eg by producer gas or residual synthesis gas. The cost of this external heating largely determines the economic viability of the process. The blast gas usually consists of the carbonization gas, which is taken off separately, and which cannot be used directly as synthesis gas because of its tar and methane content. Sometimes part of the clear gas is added to it. The carbonization gas taken from the top of the gasifier is first freed from tar and then, after mixing with the steam, recirculated into the regenerators. Extensive methane splitting occurs at the high temperatures obtaining there (over 1000°C); deposition of soot is prevented by the steam. The hot blast gas is then passed into the gasification chamber and flows through the fuel bed, giving up its heat. As in other processes, here too the fixed bed contains separate drying, carbonization and gasification zones.

Industrial-scale hot-blast gas generators use almost exclusively lignite briquettes. The extent to which complete gasification is achieved varies with the process. The bulk of the water gas produced is taken off directly above the reaction zone. The remainder serves to carbonize and dry the briquettes.

These generators are essentially restricted to the use of reactive fuels such as lignites. The heat needed to heat up the fuel and to cause it to react is conveyed to the fuel bed by the blast gas. Of the total heat carried by the blast gas, only that fraction can be used which is at the high temperature level between the temperature at the inlet to the gasifier and the final temperature at which the reaction of the steam ceases. The inlet temperature is limited by, for example, the softening temperature of the ash or, in the case of high-melting ashes, by the highest temperature tolerated by the brick lining of the regenerator. The final reaction temperature, on the other hand, depends only on the fuel. Thus, if with a less reactive fuel there is only a small difference between the inlet temperature and the final reaction temperature, to compensate for this the circulation of the hot blast gas has to be stepped up considerably if an adequate conversion is to be attained.

The circulation of large volumes of hot gas requires recuperators of correspondingly large dimensions, and implies increased losses of sensible heat. Furthermore, the reduced concentration of steam in the blast gas leads to low reaction velocities. Because of these disadvantages surrounding the use of less reactive fuels, gasification by the hot-blast gas process has been practically confined to highly reactive lignites. On the other hand, it is also manifestly difficult to achieve flexibility in adapting to fluctuations in load and regulation of gasification temperatures and throughputs.

As an example of the operating characteristics of hot-blast gas processes, an account is given below of the Pintsch-Hillebrand process. Other processes will be mentioned only briefly.

Pintsch-Hillebrand synthesis gas process

I. Gasification system

Allothermic hot-blast gas counter-current gasification (with external heating) of reactive fuels with steam in a stationary charge at atmospheric pressure.

II. Performance

A large installation with 11 generators was built before the war for the Union Rheinische Braunkohlen Kraftstoff AG hydrogenation plant at Wesseling, and was operated until the beginning of the 1960s. Other generators were operated in municipal gasworks. The data given refer to the operating figures for one generator in the installation at Wesseling.

A generator with an internal diameter of 5.36 m and external diameter of 5.80 m and a gasification area of 9 m^2 produced 5500 m^3(iN)/h of water gas, corresponding to 62 GJ/h. This represented the conversion of 3 t/h of lignite coke.

Normal load = 50 t/day

Full load = 72 t/day

III. Feed coal requirements

a. Rank: Lignites; in other generators peat briquettes also.
Moisture content 13-18%.

b. Particle size: Not stated

c. Ash: Softening point not below 1000°C; ash content 5%.

d. Coal pretreatment: Briquetting is necessary for lignites, which have a relatively low ash content.

IV. Raw gas

Carbonization gas and clear gas are produced simultaneously. The carbonization gas is charged with steam and used for gasification.

Composition:

	A. Carbonization gas	B. Water gas (clear gas)
H_2	48.0 vol %	56.0 vol %
CO	26.0 vol %	28.0 vol %
CO_2	17.5 vol %	14.0 vol %
CH_4	6.5 vol %	1.0 vol %
N_2	1.2 vol %	1.0 vol %
C_nH_m	0.8 vol %	- - vol %

Calorific value of water gas: 11.28 MJ/m^3(iN)
(2694 $kcal/m^3$(iN)

V. Description

Coal supply: The lignite is briquetted and stored at the top of the gasifier. From there it is fed into the carbonization shaft.

Carbonization: Carbonization takes place in a conical carbonization space with part of the gasifying medium. The gas enters the carbonization zone at about 700°C and leaves it at 100°C.

Gasification: The fuel passes from the conical carbonization space to the ring-shaped gasification compartment, where it is completely gasified by circulating gas and steam which have been preheated to 1200-1300°C in a regenerator. Part of the clear gas is taken off at 700°C above the gasification zone; the remainder flows into the carbonization zone.

Regenerator: Each gasifier requires two regenerators, which are heated alternately for periods of 16 minutes. One regenerator is heated to 1400-1600°C by burning a mixture of air and foreign gas, while the second regenerator gives up its heat to the purified product gas which is being recycled. The foreign gas may be part of the product gas, but it has proved better to use gas from a second gas producer (a revolving-grate generator). The ash is removed continuously through a quench chamber.

Gas scrubbing: Before being heated in the regenerator the carbonization gas (at 100°C) passes through a tar separator. The clear gas (at 700°C) passes first through a waste heat boiler for producing steam, and then (at 230°C) through a cooler in which it is charged with water vapour. It then enters an electrostatic dust and tar precipitator, after which it is either mixed with the carbonization gas or purified further for use as synthetis gas. The raw gas contains

	about 0.5 g sulphur/m^3(iN), of which 0.1 g is in organic compound form. The ash, which is rich in lime, binds a large proportion of the sulphur (cf (1, 2)).
Dust separator:	Cyclone, electrostatic precipitator
Tar separator:	Electrostatic precipitator

VI. Production of gasifying medium

The gasifying medium consists of a mixture of steam, synthesis gas and carbonization gas; hydrogen is produced by the reaction of the steam with the coal and by the thermal decomposition of the carbonization gas in the regenerator. No soot is formed, because of the presence of steam.

Heating of the regenerators is effected by the separate combustion of producer gas with air. A second gasification plant must therefore be available.

The steam produced in the waste heat boiler is sufficient for the requirements of the entire plant, including the turbines for the hot blast gas and the blowers. The clear gas is saturated with water vapour in the gas scrubbing tower.

VII. Technical data

a. Carbon conversion efficiency:	Not stated (complete gasification)
b. Gasification efficiency:	69.2%
c. Thermal efficiency:	75.3%

d. % steam decomposition: Not stated

e. Inputs:

Producer gas (for heating regenerators) (H_o = 1460 kcal/m^3(iN)):	0.56 m^3(iN)/m^3(iN) water gas 49.2 m^3(iN)/GJ water gas
Circulating hot-blast gas (= 70-90% carbonization gas, 10-30% clear gas):	0.74 m^3(iN)/m^3(iN) water gas
Steam (40% of the H_2O requirements are met by saturating the circulating gas):	0.78 kg/m^3(iN) water gas 0.069 t/GJ water gas
Additional steam requirements:	0.125 kg/m^3(iN) water gas 11.1 kg/GJ water gas
Lignite briquettes:	0.56 kg/m^3(iN) 49.2 kg/GJ water gas
Consumption is thus:	1.626 t steam/t coal (briquettes)
1000 m^3(iN)	foreign gas/t coal (briquettes)

f. Steam/oxygen ratio: Not applicable

g. Products:

Water gas yield:	1900 m^3(iN)/t coal(briquettes)
Tar:	49 kg/t coal (briquettes)

h. Pressure and temperature:

Pressure:	Slight excess pressure	
Temperature:	regenerator:	1300°C
	gasification zone:	1200-700°C
	carbonization zone:	700-100°C
Exit gas temperature:	clear gas:	700°C
	carbonization gas:	100-110°C

i. Water consumption: Not stated

j. Energy and coal consumption:

for gasification:	3 t coal (briquettes)/h
	3000 m^3(iN)producer gas/h
for energy:	4 kW/100 m^3 water gas

k. Gas purification:

Carbonization gas:	Separation of dust and tar
Clear gas:	Pre-scrubbing, separation of dust and tar, sulphur removal

VIII. Construction and development

Julius Pintsch AG

Operators

(1928) Hamburg Gasworks
(1938) Union Rheinische Braunkohlen Kraftstoff AG, Wesseling

References

(1) F. Stief, Gas- und Wasserfach 75 (1932), 581

(2) F. Stief, Gas- und Wasserfach 83 (1940), 1

(3) H. Müller, Gas- und Wasserfach 78 (1935), 431

(4) BIOS Overall Rep. No 1, p 22 (1947)

(5) F. Domann, Gas- und Wasserfach 91 (1950), 161

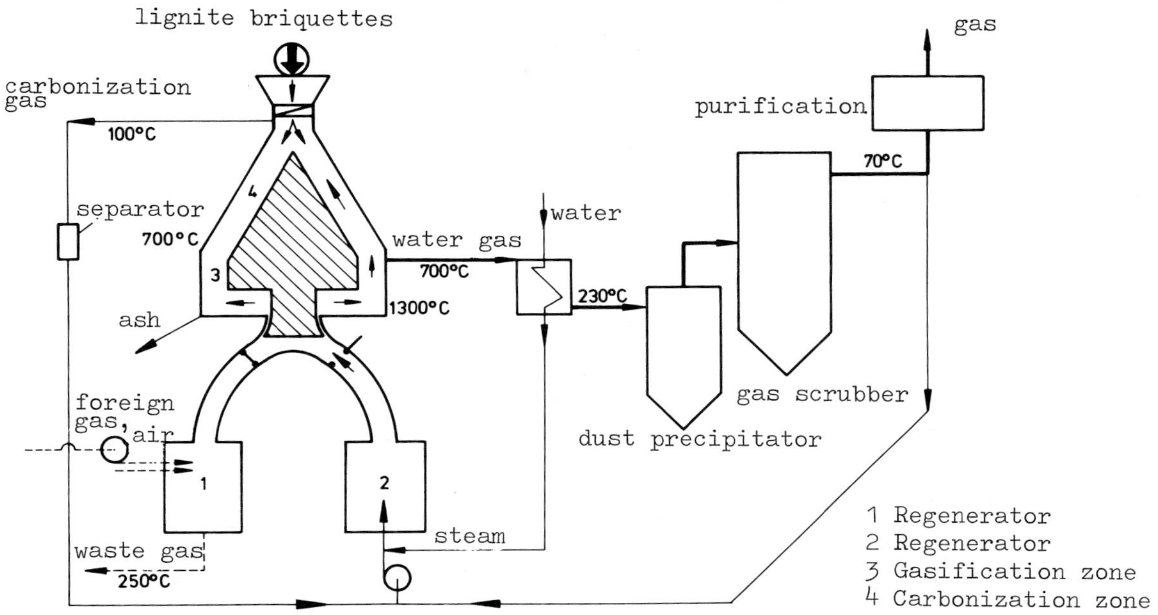

Figure 14 Diagram of the Pintsch-Hillebrand process

Didier-Bubiag process

Another example of external heating is the Didier-Bubiag process. It is used in particular for the carbonization and partial gasification (with steam) of lignite briquettes and peat in a chamber oven. The coke produced is burned with air and provides a hot flue gas for heating the chambers. The synthesis gas produced (1-2% CH_4, 80% H_2 + CO) has a calorific value of 16.75-18.0 MJ/m^3 (iN) (4000-4300 kcal/m^3(iN)).

The process was used industrially for making synthesis gas, but could not be continued on account of its low specific yield and the high cost of the plant.

References

(1) G. Duchène and R.M. Martin, Chaleur et Ind. 29 (1948), 113, 149

(2) A. Thau, Brennstoffchemie 16 (1935), 61

(3) A. Thau, Öl und Kohle 18 (1942), 721

Koppers hot-blast gas process

Another hot-blast gas process was developed by Koppers in which high-grade lignite briquettes are converted to synthesis gas by means of strongly heated circulating blast gas containing steam. Gasification is not as complete as in the Pintsch-Hillebrand process, and there is a residue of char. It differs in structural design from the Pintsch-Hillebrand process in that the regenerators stand beside the gas producers. An installation at the VEB synthesis works at Schwarzheide uses this process.

References

R. Rammler, Freiberger Forsch.-Heft A 99 (1958), 9

Tube furnace process

An allothermic gasification process for lignite has been proposed recently in which the coal, with its inherent water, was to be converted in an externally heated tube furnace. The external heating was to be effected either by combustion of part of the product gas, or by separate gasification of the residual char and combustion of the gas resulting from this, or by introducing heat from external sources, such as a high-temperature nuclear reactor. Gasification with the water still contained in the fuel would have the advantage that preliminary drying and other pretreatment of the coal would be eliminated, or at least very much simplified, and that the addition of steam of feed water quality would be unnecessary.

This process is not being developed further.

References

(1) W. Wenzel, M. Meraikib and F.H. Franke, Chem. Ing. Techn. 42 (1970), 638

(2) F.H. Franke, W. Wenzel, M. Meraikib and Chr. Beckervordersandforth, Brennstoff Wärme Kraft 25 (1973), 219

(3) F.H. Franke, W. Wenzel and Chr. Beckervordersandforth, Braunkohle 26 (1974), 173

(4) F.H. Franke, E. Pattas and K.J. Klöcker, Braunkohle 27 (1975), 362

2. Slagging-bed generator processes

The construction of slagging-bed generators resembles that of blast furnaces. In summary they can be described as unpressurized, grate-less, fixed-bed gasifiers in which the slag is tapped off in a molten state. The industrial introduction of these gas generators took place in the first few decades of this century. In Germany, gas producers of this type were operated up to the 1960s.

Their construction is basically very simple. The generator consists of a cylindrical shaft without any fittings, which narrows into a cone at its lower end and passes into the so-called hearth, in which the liquid slag collects before removal. The shaft is either brick-lined throughout or its lower part may be fitted with a water jacket for cooling. In many versions cooling is also effected by spraying the outside of the generator jacket with water. The gasifying medium is introduced directly above the layer of slag through nozzles distributed round the circumference of the gasifier, the so-called tuyères. The coal is fed in at the top of the generator. The slag is tapped off periodically from the base of the generator. As the slag flows out it is quenched with water and granulated.

Either coal or coke can be used as fuel, and the gasifying medium can be either air, air enriched with oxygen, or oxygen.

The counter-current flow of the gasifying medium and the coal results in the formation of successively layered zones up the height of the charge in which slagging of the ash, gasification, carbonization and drying of the fuel take place separately. The use of air as the gasifying medium and coal as the fuel gives rise to a number of basic problems. The gasification zone becomes larger, and the resulting gas enters the carbonization zone at a very high temperature. The tar produced there is extensively cracked and is highly viscous. The raw gas leaves the charge at a temperature of 800-900°C. As well as tar, it carries out with it a relatively large quantity of fine dust, which is formed by breakdown of the pieces of input coal as a result of the high carbonization temperatures: its entrainment is favoured by the high gas velocities needed to obtain satisfactory conversion, which are further increased by the volume increase of the reaction and by the high temperature. The high dust content and temperature of the raw gas aggravate corrosion of the parts of the plant through which it then passes, and the high temperature also necessitates a relatively large consumption of cooling water and large-diameter pipework. Utilization of the waste heat is hindered by the content of tar and dust. These problems are mitigated when coke is used.

When operating with air it is not possible to add large quantities of steam to the gasifying medium, since this would lower the temperature unduly. As a consequence of the high gasification temperature, the gas has a high content of carbon monoxide, but contains little hydrogen. Its calorific value is between 4 and 5 MJ/m^3 (iN) (1000-1100 kcal/m^3 (iN)). Air-blown slagging-bed generators are also somewhat inflexible as regards adapting to rapid changes of load. The load limits are rather narrow. A reduced load impairs the flow of slag, so that fluxes must be added.

The disadvantages of operating with air can be greatly reduced by enriching it with oxygen, or by using oxygen as the gasifying medium. The temperature in the combustion zone is then considerably higher, so that more steam can be added. The exit gas temperature drops to 400-500°C, since a smaller amount of gas is used to carbonize the coal. The carbonization of the coal is less drastic, less dust is produced, and the cracking of the tar is avoided to a large extent. There is more flexibility in adapting to fluctuating loads when oxygen is used.

There are, however, other fundamental disadvantages in the operation of slagging-bed generators. The high temperatures of the combustion and gasification zones, which are needed to ensure high reaction velocities and an almost complete burn-up of carbon, also result in large losses of sensible heat. These include increased radiation losses, the sensible heat of the raw gas and the heat content of the slag. Under some conditions there can be considerable losses of carbon as suspended dust, especially when coal is used.

All of these factors combine to keep the gasification efficiency quite low, and it barely reaches 70%, especially when air is used. On the other hand, slagging-bed generators offer a number of advantages. In the first place, they are of very simple construction and therefore relatively inexpensive. Secondly, the fuels used can include a wide range of different coals irrespective of the ash melting point, in addition to coke. However, the most significant advantage is the fact that, because of the high temperature, an extremely high shaft loading of more than 1 t/m^2 h is possible. Some generators have reached a daily throughput of 500 t of coal.

A detailed description of the BASF slagging-bed generator process is given below as an example. Literature references for other processes are also listed.

BASF slagging-bed generator process

I. Gasification system

Autothermic gasification of coke with oxygen and steam in a grate-less shaft generator. Slagging fluxes are added and the liquid slag is tapped off periodically. Part of the carbon to be gasified may be added in the form of liquid or gaseous hydrocarbons. Air and air enriched with oxygen have also been tested successfully as gasifying agents.

II. Performance

Generators of this type were developed and used for the commercial production of synthesis gas by BASF from 1945 onwards; they had a shaft diameter of 4.8 m (hearth diameter 3.8 m) and an overall height of 9 m (depth of coke charge 4.5 m). A coke throughput of up to 24 t/h was attained, with a synthesis gas yield of up to 55 000 m^3 (iN)/h. The following figures give the operating data for these generators:

specific throughput of coke: 1.33 t/m^2 h

specific gas production: 3040 m^3 (iN)/m^2 h

(both values are expressed in terms of shaft cross-section)

III. Fuel requirements

a. Type:	(A) Coke from hard coal (87% C) (B) Coke (87% C) + 40% light benzine
b. Particle size:	Lump coke (particle-size distribution not stated)
c. Ash:	Not stated

d. Pretreatment: Not stated

IV. Raw gas composition and calorific value (dry basis)

	(A)	(B)
H_2	31 vol %	41 vol %
CO	62 vol %	53 vol %
CO_2	6 vol %	5 vol %
CH_4	--------	0.1 vol %
N_2	1 vol %	1 vol %
Sulphur compounds:	not stated	not stated
Calorific value:	11.75 MJ/m^3(iN) (= 2807 kcal/m^3(iN)	11.94 MJ/m^3(iN) (= 2850 kcal/m^3(iN))

V. Description of process

The coke is mixed with limestone (in order to improve the fluidity of the slag) and is kept in a storage bunker. It is fed into the generator through a lock-hopper by a carefully automated charging system triggered with a sensing arm which monitors the height of the charge in the generator.

Oxygen (and in some cases air) for gasification is saturated with water vapour in a saturator, and more steam is added at the outlet of the saturator. The gasifying mixture, containing at least 40 vol % oxygen, is then preheated to 200-300°C in a steam heater. Light benzine containing a fixed amount of steam is passed in parallel through a vaporizer and then also heated to about 250°C. The gasifying medium and the benzine vapour are premixed in the injection nozzles of the reactor and then blown into the gasification zone at a slight angle to the base of the generator. This procedure ensures a satisfactory flow of slag on combustion.

The molten slag is periodically run off through tap-holes in the base and granulated by a jet of water. The granules are subsequently separated from the water in a settling tank.

The gas produced in the generator passes first through the dry-dust removal unit (dust collector and two cyclones in parallel). This coke dust is returned to the generator through two of the twelve tuyères. The dust removal unit is followed by a waste heat boiler. The high-temperature stage of this boiler produces heating steam for the benzine vaporizer and for pre-heating the gasifying medium, while the second stage gives process steam at 15 bar.

The gas, now cooled to about 220°C, is freed from the remaining dust in a scrubber-cooler and a Theisen disintegrator. The data available make no reference to desulphurization.

No information is given about behaviour under fluctuating loads, running at partial load, or conditions for start-up and shutdown.

VI. Production of gasifying medium

Oxygen:	Linde-Fränkl air fractionation plant
Steam:	No information: the low-pressure steam required was presumably taken from the works steam line.

VII. Technical data

	(A)	(B)
a. Carbon conversion efficiency:	Not stated	
b. Gasification efficiency:	(94%)	(92%)

	(A)	(B)
c. Thermal efficiency:	(91%)	(91%)

The efficiencies have been estimated from published data, but the figures available are insufficient for an accurate calculation.

d. % steam decomposition: Not stated

	(A)	(B)
e. Inputs:	0.44	0.265 kg/m^3(iN) raw gas
Coke:	37	22 kg/GJ raw gas
Light benzine:	-----	0.117 kg/m^3(iN) raw gas
		9.8 kg/GJ raw gas
Steam:	0.363	0.28 kg/m^3(iN) raw gas
	30.8	23.5 kg/GJ
	0.824	kg/kg coke
Oxygen (98%):	0.223	0.226 m^3(iN)/m^3(iN) raw gas
	19.0	18.9 m^3(iN)/GJ raw gas
	0.507	m^3(iN)/kg coke
Flux (limestone):	Not stated	
f. Steam/oxygen ratio:	1.63	1.24 kg/m^3(iN)
g. Products:		
Gas yield:	2273	m^3(iN) raw gas/t coke
Steam production (15 bar):	0.186	0.218 kg/m^3(iN) raw gas
	15.8	18.3 kg/GJ raw gas
	0.423	kg/kg coke
h. Gasification pressure:	Atmospheric	
Gasification temperature:	Not stated	
Exit gas temperature:	630°C	760°C

	(A)	(B)
i. Water consumption:	Not stated	
j. Other energy consumption:	Not stated	

k. Gas purification: From the generator the gas passes into the dry-dust removal unit, in which the greater part of the entrained coke dust is removed. This dust consists mainly of particles between 0.05 and 1.0 mm in diameter. About 8% of the particles are larger than this and 15% smaller. Because of the decreased velocity the dust collector catches all particles larger than 0.2 mm: no information is available on the efficiency of the cyclones. Data are also lacking for the efficiency of the scrubber-cooler and the Theisen disintegrator. No description is given of the desulphurization of the raw gas.

VIII. Construction and development

BASF AG, Ludwigshafen

References

F. Duftschmid and F. Markert, Chem. Ing. Techn. 32 (1960), 806

Thyssen-Galoczy process

W. Demann, Brennstoffchemie 30 (1949), 187

L.L. Newman, Ind. Eng. Chem. 40 (4) (1948), 578

Leuna slagging-bed generator process

F. Sabel, Brit. Eng. Obj. Sub-Committee Report No 199 (1945)

F. Sabel, Gas- und Wasserfach, Gas 92 (1951), 273

3. British Gas Corporation high-pressure slagging gasifier process

I. Gasification system

Autothermic counter-current pressure gasification of lump coal or coke with steam and oxygen in a fixed-bed reactor. In contrast to the conventional Lurgi process for pressure gasification, the temperature in the combustion zone is raised above the melting point of the ash.

II. Performance

During the last 20 years in the UK both the British Coal Utilization Research Association (BCURA) and the Gas Council have experimented with slagging gasifiers which can be regarded as further developments of the Lurgi process. BCURA operated an experimental gasifier with a throughput of up to 0.9 t/h which employed only a slight excess pressure, and also some smaller research gasifiers at atmospheric pressure, while the Gas Council used a rebuilt Lurgi gasifier with a throughput of 5.7 t/h and an internal diameter of 0.91 m. This converted gasifier was equipped with a brick lining to catch the liquid slag, a quench tank and a slag lockhopper. Some fairly early data (1965) for this generator, which was only run experimentally, are given below. The researches of the Gas Council are being continued by the British Gas Corporation, who are operating at Westfield in Scotland an experimental plant with three conventional Lurgi gasifiers of 2.80 m diameter and one gasifier converted to slagging operation. No data are available for this plant.

The US Bureau of Mines has also carried out experiments on a pilot scale for gasifying lignite in a Lurgi gasifier converted to molten slag removal. No data are available on these experiments either, which appear to have been abandoned.

Gas Council experimental plant:

Diameter of generator:	0.91 m
Coal throughput:	up to 5.7 t/h
Gas production:	up to 8000 m^3(iN)/h
Specific throughput of coal:	up to 8.7 t/m^2 h
Specific gas production:	up to 12 300 m^3(iN)/m^2 h

III. Feed coal requirements

a. Rank: Most of the experiments were carried out with coke, but weakly or moderately caking bituminous coals were also used. The data given here refer to Donisthorpe coal with the following composition (dry basis):

C	71.3	wt %
H	5.0	wt %
O	13.2	wt %
N	1.55	wt %
S	1.45	wt %
Ash	7.35	wt %

Calorific value H_o = 29.6 MJ/kg (7053 kcal/kg)
Moisture content of feed coal 12.7 wt %

b. Particle size: The great majority of experiments were carried out with coal which had been fairly closely graded to a particle size of 25-37 mm. However, it was found that there was no significant effect on performance when coal of particle size 12-25 mm was used. The following data refer to the size range 25-37 mm.

c. Ash: Since the ash is melted in the reactor, the thermal efficiency is dependent to some extent on the ash content of the coal. The melting point of the ash is not in general critical, though in most experiments dolomite or blast furnace slag had to be added as a flux. For Donisthorpe coal the melting point of the ash was about 1350 °C. In the experiments described the ratio by weight of coal ash to blast furnace slag was 4 : 3.

d. Coal pretreatment: The pretreatment of the coal appears to have been essentially the same as that required for the conventional Lurgi process, ie crushing, grading and, if necessary, drying. No information is given on the maximum acceptable content of moisture or dross in the coal.

IV. Raw gas composition (dry, tar-free basis)

Depending on experimental conditions, the raw gas actually contains varying amounts of nitrogen (between about 5 and 7%). This comes from the combustion gas of the taphole burner, which is fuelled with town gas and air enriched with oxygen. However, the published gas analyses are calculated for a nitrogen-free basis for the purposes of comparison.

	Coke	Donisthorpe coal
H_2	27.25 vol %	28.05 vol %
CO	70.35 "	61.2 "
CO_2	1.85 "	2.55 "
CH_4	0.45 "	6.71 "
C_2H_6	0.0 "	0.94 "
Calorific value:	12.50 MJ/m^3(iN) (= 2987 $kcal/m^3$(iN))	14.61 MJ/m^3(iN) (= 3491 $kcal/m^3$(iN))

V. Description

The reactor differs from the conventional Lurgi pressure gasifier only in its lower section. Steam and oxygen are blown through four water-cooled tuyères into the brick-lined hearth. The problem of removing the slag does not appear to have been entirely solved. Tests have been carried out both with periodic tapping and with continuous flow. In order to obtain a satisfactory flow of slag the discharge channel from the generator has to be additionally heated by a burner. The slag falls into a pressurized water bath beneath the reactor, where it is solidified and granulated. The solidified lumps are removed from the pressurized bath through a lockhopper system.

The upper section of the gasifier consists of the usual steel pressure vessel, cooled by water under pressure and furnished with a fuel distributor and lockhopper above.

Further treatment of the gas at Westfield is carried out in the same way as for conventional Lurgi pressure gasification.

VI. Production of gasifying medium

Oxygen: Fractionation of air by the Linde-Fränkl process. In the Gas Council's experiments at Solihull liquid oxygen was evaporated.

Steam: In an industrial installation part of the steam for gasification could be obtained from the pressurized water cooling jacket of the gasifier. The remainder has to be produced in a separate boiler. The process is not self-sufficient in steam. In the experiments, all the steam was obtained from external sources.

VII. Technical data

The data given here come from an 81-hour test run in which the weakly caking Donisthorpe coal described above was gasified. Some of the figures would probably differ in commercial operation.

a. Carbon conversion efficiency: 99.6%

b. Gasification efficiency: 90.0% (including tar, etc)

c. Thermal efficiency: 68.3%

d. % steam decomposition: 83%

e. Inputs:

Oxygen:	274	m^3(iN)/t raw coal
	0.186	m^3(iN)/m^3(iN) raw gas*
	12.8	m^3(iN)/GJ raw gas
Steam:	238	kg/t raw coal
	0.162	kg/m^3(iN) raw gas*
	11.1	kg/GJ raw gas
Raw coal:	0.680	kg/m^3(iN)raw gas*
	46.6	kg/GJ raw gas

f. Steam/oxygen ratio: 0.88 kg/m^3(iN)

g. Products:

Gas yield:*	1471	m^3(iN)/t raw coal
Tar, etc:	59.3	kg/t raw coal

h. Gasification pressure: 20.7 bar

Gasification temperature: 1250-1500°C

Exit gas temperature: 350-450°C

State of gasifying medium:

Oxygen:	250°C;	103 bar
Steam:	450°C;	37.9 bar

---

*The volumes of raw gas again refer to nitrogen-free gas.

i. Water consumption: Not stated

j. Other energy consumption:

The following quantities are used for fuelling the taphole burner:

Town gas:	20.7	m^3(iN)/t raw coal
(4221 kcal/m^3(iN))	≙ 366	MJ/t raw coal
	0.014	m^3(iN)/m^3(iN) raw gas
	0.96	m^3(iN)/GJ raw gas
Oxygen:	6.22	m^3(iN)/t raw coal
	4.23	m^3(iN)/10^3 m^3(iN) raw gas
	0.29	m^3(iN)/GJ raw gas
Air:	111	m^3(iN)/t raw coal
	75.5	m^3(iN)/10^3 m^3(iN) raw gas
	5.2	m^3(iN)/GJ raw gas

Electricity consumption: Not stated

k. Gas purification: The raw gas leaving the generator is first scrubbed with condensate in a quenching cooler, and then passes successively through a waste heat boiler and a final cooler for condensing out tar, etc, and gas liquor. The amount of gas liquor is 187.8 kg/t raw coal (≙ 8.7 kg/GJ raw gas) and is therefore considerably less than in the conventional Lurgi process, because of the increased steam decomposition. The amount of suspended dust is 9.04 kg/t raw coal. No attempt was made to reduce the amount of dust by returning the tar to the generator, though this would be considered in a commercial installation.

VIII. Construction and development

British Gas Corporation, Westfield, Scotland;

formerly Gas Council, Solihull, England

References

(1) D. Hebden, J.A. Lacey and A.G. Horsler,
J. Inst. Gas Eng. 5 (1965), 367

(2) D. Hebden and R.F. Edge,
Trans. Inst. Gas Eng. 108 (1958-59), 492

(3) Coke and Gas, 21 (1959), 13

(4) Arch.f. Energiewirtschaft, 15 (1959), 600

(5) D. Hebden,
Gas World, 178 (1973), 341

(6) The Westfield Development Centre,
booklet published by the British Gas Corporation, 1975

(7) J.A. Gray,
Coal Energy Quarterly, No 11 (1976), pp 15-22

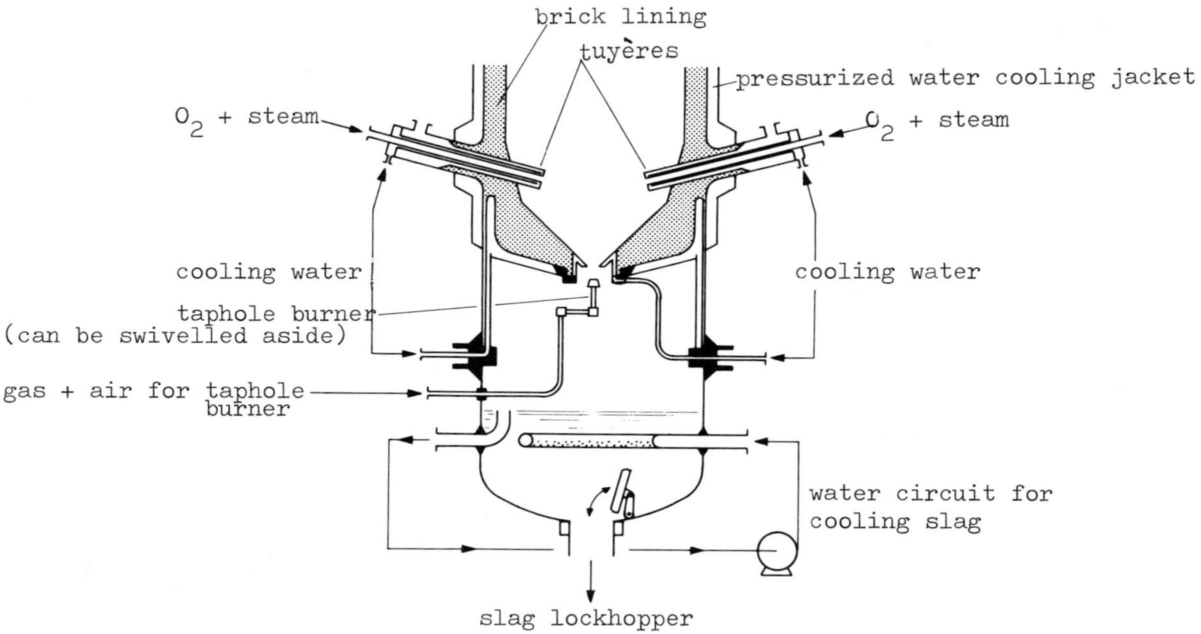

<u>Figure 15</u> High-pressure slagging gasifier (lower section)
(British Gas Council, Solihull, England)

4. Hygas process

I. Gasification system

Combined hydrogasification and steam gasification in a multistage reactor. In the first stage mainly the volatile constituents of the pulverized coal in suspension are partially gasified and dehydrated. The second stage involves hydrogasification in a fluidized bed with a hydrogen-rich mixture of gas and steam. The residual char is then gasified with steam in order to produce process hydrogen: three alternative methods are being developed for this last stage.

II. Performance

There exists at present a pilot plant with a coal throughput of 3 t/h. Full performance has only been achieved in short runs, with the hydrogenation section operating separately and not directly coupled to the char gasification stage. No exact figures have been published on the dimensions of the reactor or the yield of raw gas. It has merely been reported that the pilot plant is designed for an SNG output of 1670 m^3(iN)/h. The gasification of peat has recently been investigated.

The data given below are based on the design data published by the Institute of Gas Technology (IGT) for large-scale installations to produce 310 000 m^3(iN) SNG/h, equivalent to 10 800 GJ/h (= 2620 Gcal/h). Directly measured values were not available.

III. Feed coal requirements

a. Rank: All lignites and non-anthracitic coals. Tests in 1976 were with Illinois No 6 (40% volatile constituents (dry basis), strongly caking) and Rosebud coal (48% volatile constituents (dry basis), non-caking).

b. Particle size: Dust with a fairly flat particle-size distribution between 45 μm and 850 μm: not more than about 1-2% may exceed 1.5 mm.

c. Ash: Not stated

d. Coal pre-treatment: After drying and grading, caking and swelling coals (eg Illinois No 6) are first burned off to about 10% and pre-oxidized in a preliminary reactor (fluidized bed) at 350-400°C to eliminate the tendency to cake. This produces steam and a combustible heating gas which can be used as boiler fuel. Lignite and non-caking hard coals do not require preliminary oxidation. It is intended to gasify all coals without preoxidation at a later stage.

The coal dust is mixed with a light aromatic oil (produced in the process itself) to form a slurry, which is pumped into the gasification reactor. The slurry may contain up to 45 wt % coal dust.

IV. Raw gas composition and calorific value (dry, oil-free basis)

There are three different sets of figures, depending on the method used for gasifying the residual char (for hydrogen production):

	Steam-oxygen process	Steam-iron process	Electrothermal process
CO	23.8 vol %	11.0 vol %	25.7 vol %
CO_2	24.5 vol %	10.6 vol %	17.4 vol %
H_2	30.2 vol %	33.6 vol %	29.2 vol %
CH_4	18.6 vol %	29.0 vol %	24.0 vol %
C_2H_6	0.6 vol %	1.5 vol %	0.9 vol %
H_2S	1.2 vol %	2.2 vol %	1.6 vol %
other	1.1 vol %	2.1 vol %	1.2 vol %
Calorific value:	H_o = 15.0 MJ/m^3(iN) (≙ 3582 kcal/m^3(iN))	H_o = 18.8 MJ/m^3(iN) (≙ 4487 kcal/m^3(iN))	H_o = 17.57 MJ/m^3(iN) (≙ 4194 kcal/m^3(iN))

V. Description

Either after being burned off and preoxidized or directly after drying and grinding, the feed coal is slurried with a light oil and injected into the reactor by slurry pumps. Preoxidation causes losses of about 20%. In the reactor the oil is vaporized by the product gas flowing counter-current and the coal is dried further. It then moves down into the first gasification stage, where it is partially gasified and dehydrated by the gas from the second gasification stage. The second hydrogasification stage lies below the first, and the coal reaches it by overflow from the first stage. The first stage is constructed as a narrow tubular section, in which the coal is entrained by the rising current of gas. This stage is followed by a wider section, in which the coal particles can settle and overflow into the second stage, which is in the form of a fluidized bed. A hydrogen-rich gas/steam mixture is passed up through the fluidized bed from jets in its base, and there is a simple overflow for the residual char, which falls into the hydrogen producer below. The latter may be any one of the following three versions:

a. Gasification is effected by a steam-oxygen mixture in a stationary charge under pressure. The ash falls into a water lock chamber and is removed hydraulically. The gas produced is passed directly to the fluidized bed of the second hydrogasification stage. The dust removed at the top of the entire installation (ie beyond the preheater and oil vaporizer) is returned to this stage via multiclones. Since the temperature in this stage (so far tested only with lignite) is about 1050°C, there is no formation of slag or clinker, but a sintered ash with almost spherical particles of uniform size is obtained. Carbon conversion efficiencies of more than 98% have been achieved.

b. In the steam-iron process the residual char from the second hydrogasification stage is treated with air and steam to give producer gas, which in the reduction stage of the steam-iron process reduces iron oxide (siderite, 70-360 μm) to the metal. The metal is then reoxidized with steam in the oxidation stage, and returned to the reduction stage. The mixture of hydrogen and steam produced is passed through the fluidized bed of the hydrogasifier. The oxidation and reduction of the iron oxide take place at 800-850°C, and the pressure is about 75-80 bar, as in other parts of the plant. Since October 1975 the steam-iron process has been tested over a long period in a pilot plant with caking coal as the fuel.

c. In the electrothermal process the residual char from hydrogasification is gasified with steam at 1050-1100°C in a fluidized bed heated directly by passing direct current through it. Since the electrical resistance of the fluidized bed fluctuates widely, electronic controls have been developed to keep the heating approximately constant. Because of the steep rise in the cost of electricity, further development of this method has been deferred for the time being. The long-term intention is to produce the electricity required by burning some of the residual char itself. This process is the simplest of the three variants, but it is also the most expensive.

The raw gas leaves the reactor upstream of the preheating and oil vaporizing stage at a temperature of 300°C. In addition to the components already listed it contains pasting oil, water vapour, coal dust and traces of ammonia, hydrogen cyanide, etc. In the pilot plant mentioned, further treatment consists in removing oil, dust and water-soluble trace impurities from the gas by scrubbing with water in a scrubber-cooler. The oil and the water are then separated before subsequent treatment.

The gas is freed from acid gases in conventional equipment. In the pilot plant a water/diglycolamine mixture is used for scrubbing, in order to cope with gas compositions varying over a wide range, but optimized processes would be used in an industrial plant. The wash solution used in the spray tower is regenerated with steam at atmospheric pressure in a second tower and then recirculated. The CO_2-H_2S mixture is desulphurized in a Claus kiln. The amine wash is followed by scrubbing with alkali to free the gas from residual impurities, especially sulphur compounds, which would affect the nickel methanation catalysts.

In the methanation stage the temperature must be kept strictly above 230°C (to prevent formation of $Ni(CO)_4$) and below 530°C (to prevent deposition of carbon). Since the reaction $CO + 3\ H_2 = CH_4 + H_2O$ is strongly exothermic, the gas must be diluted in order to reduce the CO content and thus keep the evolution of heat within the allowed limits. Dilution is carried out with methanated end-product gas, part of which is introduced into the circuit after cooling.

The temperature control of methanation can be improved by carrying it out in two steps, with intermediate cooling. With this procedure, to preheat the fresh gas for the first step the completely methanated gas passes through the heat exchanger and is then cooled further with condensation of the water formed in the methanation reaction. In an industrial plant the heat of reaction would be used for steam production, but this is not included in the pilot plant.

No information has been published on the dynamic behaviour of the reactor or the conditions for start-up and shutdown.

VI. Production of gasifying medium

a. Oxygen: Conventional fractionation of air

b. Steam: Possibly from methanation stage or cooling jacket. It is not stated whether the process can be made self-sufficient in steam.

VII. Technical data

a. Carbon conversion efficiency: 96-99%

b. Gasification efficiency: Not stated

c. Thermal efficiency: Not stated

d. % steam decomposition: Not stated

e. Inputs:

Steam-oxygen process:

Oxygen:	$765 \cdot 10^3$ m^3(iN)/h ≙ 1378 m^3(iN)/t feed coal ≙ 69.7 m^3(iN)/GJ in SNG product
Steam:	532 t/h ≙ 0.958 t/t feed coal ≙ 0.048 t/GJ in SNG product
Steam/oxygen ratio:	6.95 kg/m^3 (iN)
Gasified coal:	555 t/h ≙ 0.051 t/GJ in SNG product
Electricity supply:	Not stated

Steam-iron process:	No oxygen is required, since the residual char is gasified with air
Steam:	674 t/h ≙ 0.965 t/t feed coal ≙ 0.061 t/GJ in SNG product
Gasified coal:	700 t/h ≙ 0.064 t/GJ in SNG product
Electricity supply:	Not stated

Electrothermal process:	No oxygen is required, since the heating is electric
Steam:	417 t/h ≙ 0.701 t/t feed coal ≙ 0.038 t/GJ in SNG product
Gasified coal:	587 t/h ≙ 0.054 t/GJ in SNG product

Electricity required for gasifying residual char: 24.7 kW/t feed coal

≙ 1.3 kW/GJ in SNG product

Other electricity supply: Not stated

f. Steam/oxygen ratio: See VII e. above

g. Products:

Yields of raw gas: Not stated

By-products: In the first stage of hydrogasification at 650°C it is mainly the volatile constituents which are expelled from the coal and gasified and dehydrated. Some of the oil formed as one of the reaction products is recycled as pasting oil for the coal slurry, and the remainder is removed. Its composition is not stated. However, the following figures are given for the amounts of fuel obtainable in the three alternative processes:

Steam-oxygen process

704 GJ/h (= 168 Gcal/h)

≙ 1.4 GJ/t feed coal (= 0.334 Gcal/t feed coal)

2.18 MJ/m^3(iN) SNG
(= 520 $kcal/m^3$ SNG)

64.2 MJ/GJ in SNG product
(= 64.1 Mcal/Gcal in SNG product)

Steam-iron process

4810 GJ/h (= 1150 Gcal/h)
7.1 GJ/t feed coal
(= 1.7 Gcal/t feed coal)

15.6 MJ/m^3(iN) SNG
(= 3.73 $Mcal/m^3$ SNG)

0.44 GJ/GJ in SNG product
(= 441 Mcal/Gcal in SNG product)

Electrothermal process

880 GJ/h (= 210 Gcal/h)

1.50 GJ/t feed coal
(= 0.36 Gcal/t feed coal)

2.84 MJ/m^3(iN)
(= 678 $kcal/m^3$ SNG)

80.3 MJ/GJ in SNG product
(= 80.2 Mcal/Gcal in SNG product)

h. Gasification pressure: 70 bar

Gasification temperatures: 1st stage of hydrogasification 650°C

2nd stage of hydrogasification 950°C

Exit gas temperature: 300°C

i. Water consumption: Not stated

j. Coal throughput:

Steam-oxygen process: for gasification: 557 t/h

for energy production, etc: Not stated

Steam-iron process:	for gasification:	700 t/h
	for energy production, etc:	not stated
Electrothermal process:	for gasification:	587 t/h
	for producing electricity:	13.8 t/h
	other energy production:	not stated

k. Gas purification: As described in Section V, the gas is first freed from oil, dust and water-soluble trace impurities in a scrubber, and then treated further in conventional equipment. Gas of any quality required can be produced. No information is given on the content of dust or chemical impurities in the raw gas.

VIII. Construction and development

Manufacturer

Institute of Gas Technology (American Gas Association), Chicago, Illinois, USA

Operator

Same as manufacturer

References

(1) C.L. Tsaros and T.K. Subramaniam, Report for US Dept of the Interior, Office of Coal Research and American Gas Association, February 1971

(2) F. Schora, Jr, B.S. Lee and J. Huebler, Clean Fuels from Coal, Symposium, Chicago, 10-14 September 1973

(3) P.B. Tarman,
5th Synthetic Pipeline Gas Symposium, Chicago, 29-31 October 1973

(4) B.S. Lee and P.B. Tarman,
6th Synthetic Pipeline Gas Symposium, Chicago, 28-30 October 1974

(5) B.S. Lee,
8th Synthetic Pipeline Gas Symposium, Chicago, 18-20 October 1976

(6) B.S. Lee,
Oil and Gas J. <u>77</u>, 1 August (1977), 76

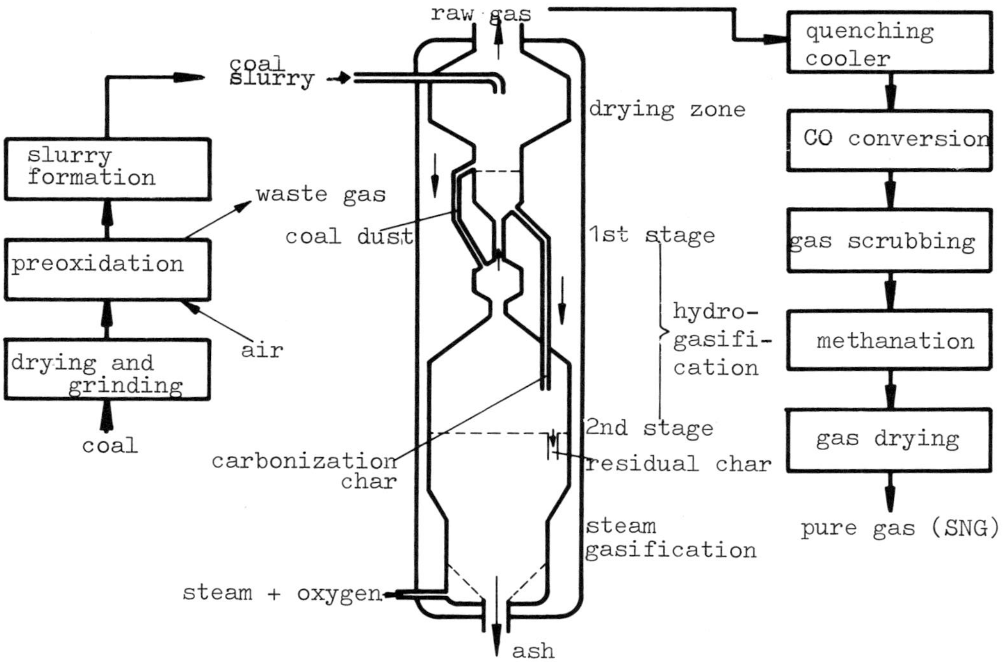

Figure 16 Hygas reactor (Institute of Gas Technology)

5. Cogas process (COED)

I. Gasification system

Allothermic process for gasifying fine char with steam in a fluidized bed. The process involves the combination of multistage fluidized-bed carbonization of fresh coal with steam gasification of the resulting char at low pressures. It is a variation derived from the COED process (Char-Oil-Energy Development Project) developed by the Food Machinery and Chemicals Corporation (FMC) in the USA.

The heat requirements are met by separate air combustion of a portion of the char. The studies are being carried out by the recently founded Cogas Development Company in the USA and under license in the UK by the British Coal Utilization Research Association (BCURA). The heat transfer agent can be either ceramic pellets (Cogas) or the fine char itself (BCURA). Carbonization yields the usual products, and gasification of the carbonization char yields synthesis gas.

II. Performance

Two experimental plants have been constructed under contractual arrangements between the Cogas Development Co and BCURA. The throughput in the UK is 4 t coal/h and in the US plant 0.2 t/h. These plants have been in operation since 1974. In the USA a demonstration plant for a daily coal throughput of about 2000 t is planned, and should commence operation in 1981.

It would appear that the raw gases obtained from the two processes have approximately the same composition. The data quoted cannot always be attributed unambiguously to one or other of the two experimental plants, though the figures given in Section VII can be identified with certainty as referring to the BCURA installation.

III. Feed coal requirements

a. Rank: Tests are described with Illinois No 6 bituminous and Wyoming sub-bituminous coal.

b. Particle size: Not stated

c. Ash: The ash is either slagged (BCURA) or deposited as suspended ash.

d. Coal pretreatment: Grading and drying

IV. Raw gas

Composition:

a. Pyrolysis gas, partially purified:	(Illinois No 6)	(Wyoming sub-bituminous)
H_2	26.2 vol %	16.6 vol %
CO	7.4 vol %	17.2 vol %
CO_2	9.3 vol %	20.1 vol %
CH_4	34.0 vol %	34.9 vol %
C_2-C_4	8.3 vol %	7.6 vol %
H_2S *)	10.8 vol %	1.2 vol %
N_2	4.0 vol %	2.4 vol %
Calorific value:	27.0	24.5 MJ/m^3(iN)
	(6460)	(5840 kcal/m^3(iN)

*)The H_2S figure is very high, possibly owing to a misprint in the original publication

b. Mixed gas from pyrolysis and gasification (purified):

	H_2	57.9	vol %
	CO	31.2	vol %
	CO_2	6.6	vol %
	CH_4	4.0	vol %
	N_2	0.3	vol %
	S	0.002	vol %

Calorific value: 12.9 MJ/m^3(iN) (3080 kcal/m^3(iN))

c. SNG (after methanation):

H_2	0.4	vol %
CO	0.7	vol %
CH_4	93.0	vol %
C_2-C_4	2.7	vol %
Inert gas	3.2	vol %

Calorific value: 37.3 MJ/m^3(iN) (8900 kcal/m^3(iN))

d. Producer gas:

H_2	49.5	vol %
CO	32.5	vol %
CO_2	15.6	vol %
N_2	2.4	vol %

V. Description

The coal is first carbonized in several stages in a series of fluidized beds by a counter current of gasifying medium. The reactive char thus produced is gasified with steam in a fluidized-bed reactor at low pressures. The heat required for the reaction is furnished by separate air combustion of the char, also in a fluidized bed. The heat is conveyed either by char itself circulating between the combustion furnace and the gas generator (BCURA), or by heat carriers - pellets made of ceramic or slag. The carbonization gas and water gas are treated in commercially available equipment.

VI. Production of gasifying medium

Steam: Not stated; probably from waste heat boilers.

VII. Technical data

In order to obtain a high efficiency, special attention has been paid to developing heat recovery in the process. Steam is obtained from the waste heat of the synthesis gas and flue gas; the air is preheated by heat exchangers, and the hot flue gas is to be passed to a turbine which drives the air compressor.

a. Carbon conversion efficiency: 96%

b. Gasification efficiency: 68%

c. Thermal efficiency: 86%

d. % steam decomposition: Not stated

e. Inputs:

Air (sensible heat)	2.5 GJ/h
Coal	18.4 GJ/h
Steam	1.7 GJ/h
Fine char + suspended dust	7.1 GJ/h

(amount of char circulating)	10.9 GJ/h	
f. Steam/oxygen ratio:	Not applicable	
g. Products (heat balance):	12.6 GJ	latent heat in the gas
	2.9 GJ	sensible heat in the gas
	3.3 GJ	latent heat in flue gas
	6.7 GJ	sensible heat in flue gas
	1.3 GJ	sensible heat in suspended dust
	6.3 GJ	sensible heat in fine char
	5.4 GJ	losses by radiation, slag and cooling

h. Gasification temperature and pressure; exit gas temperature: In the four fluidized-bed carbonizers pressures of up to 1.5 bar and temperatures of between 300 and 800°C are reached. In the fluidized-bed combustion chamber pressures of up to 3 bar and temperatures of between 800 and 900°C are used.

i. Water consumption: Not stated

j. Other consumption for energy production: Gas and oil: 2.8 GJ/h

k. Gas purification:	Commercially available equipment is to be used. The following steps are contemplated for purification and further treatment:
To produce gas of medium calorific value:	Mixing of carbonization gas and water gas, desulphurization and dissolving out CO_2 under pressure.
To produce synthesis gas for making hydrogen:	Scrubbing under pressure to remove hydrogen sulphide and CO_2.
To produce pipeline gas (SNG):	Mixing of carbonization gas and water gas, conversion of water gas, desulphurization, CO_2 scrubbing, refined desulphurization and methanation.
Processing of wastes:	The residues from the carbonization gas are processed to obtain hydrogen sulphide, phenols and ammonia. The hydrogen sulphide is combined with the sulphur deposits from desulphurization and hydrogen production and processed to produce sulphur.

VIII. Construction and development

a. Process using pellets as heat carriers

Process development

Cogas Development Company, Princeton, New Jersey, formed jointly by the following firms:

Consolidated Natural Gas Co
FMC Corp
Panhandle Eastern Pipe Line Co
Republic Steel Corp
Tennessee Gas Pipeline Co
Rocky Mountain Energy Co

Engineering work and construction of the experimental plant

Dravo Corp, Chemical Plants Division, Pittsburgh, Pennsylvania

Location of experimental plant

Princeton, New Jersey

Financial backing

The estimated cost of the projected demonstration plant which is to be built in the USA is $ 276 million. Half of this sum will be provided by DOE (formerly ERDA).

b. Process using char as the heat carrier

Process development

BCURA Ltd (National Coal Board)

Location of experimental plant

Leatherhead, England

References

(1) R. Bloom Jr, and R.T. Eddinger, Sixth AGA Synthetic Pipeline Gas Symposium, Chicago 1974

(2) L.H. Diersdorff Jr and R. Bloom Jr, Paper at Society of Automotive Engineers Meeting, August 1973

(3) R.T. Eddinger and R. Bloom Jr, Third AGA Synthetic Pipeline Gas Symposium, November 1970

(4) R.T. Eddinger, Fourth AGA Synthetic Pipeline Gas Symposium, October 1972

(5) H. Malakoff, in Chem. Engineering Progr. Coal Processing Technology, Vol 2, 1975, New York, p 58

(6) K.H. van Heek, Glückauf 113 (1977), 383

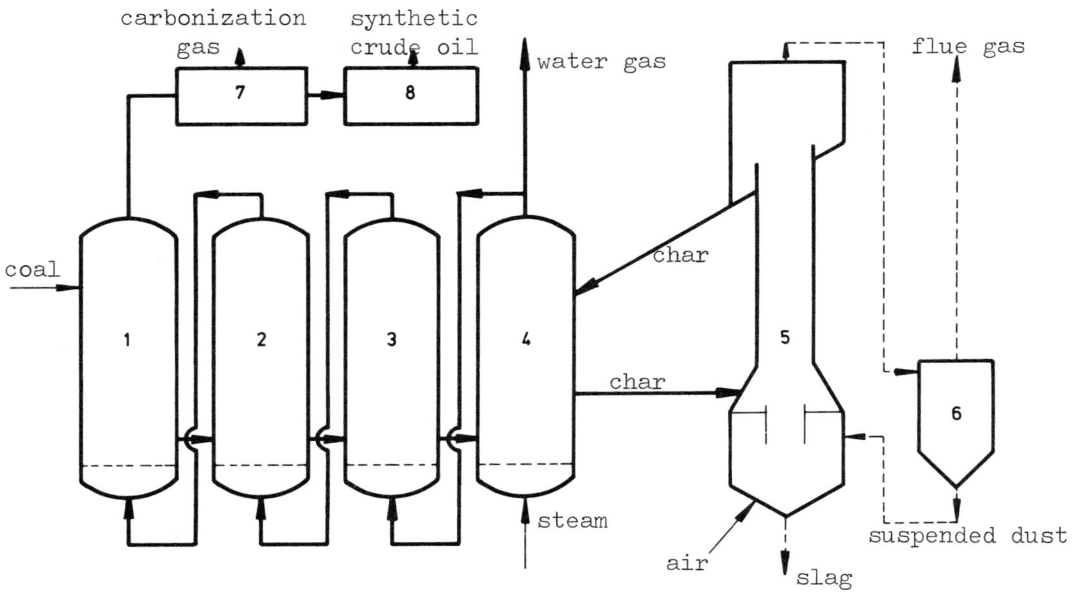

1-3 fluidized-bed carbonizers
4 fluidized-bed generator
5 fluidized-bed burner
6 suspended dust cyclone
7 oil separator
8 oil hydrogenation

<u>Figure 17</u> Cogas process with circulating heat carrier

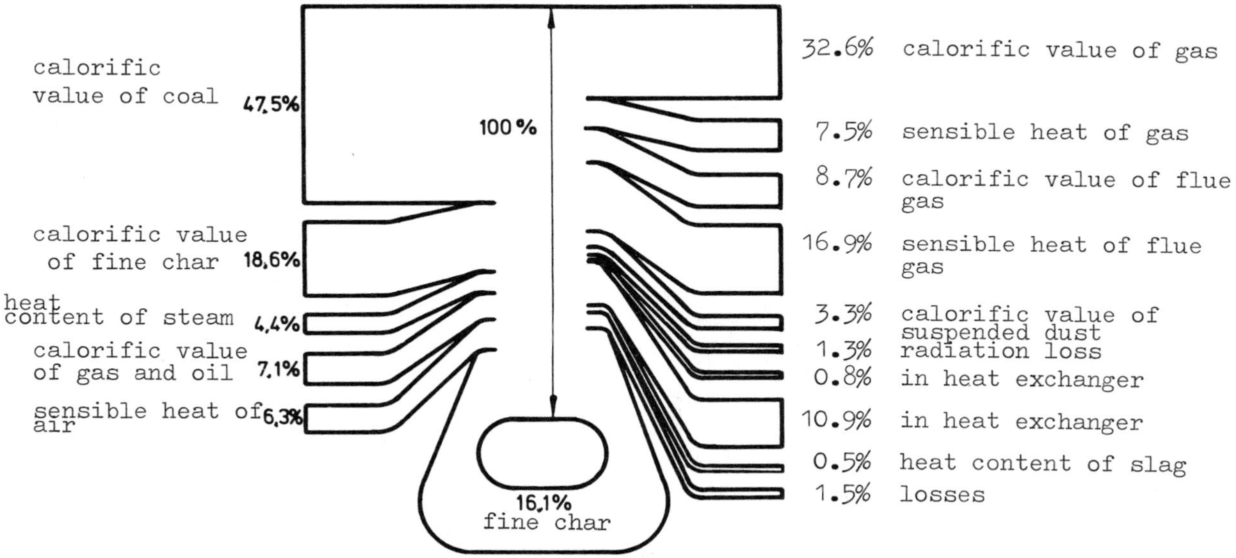

Figure 18 Heat balance for the Cogas process

6. CO_2 acceptor process

I. Gasification system

Steam gasification of coal in a fluidized bed under pressure; 75% of the heat required for the endothermic reaction is supplied by the reaction of an acceptor with CO_2:

$$CaO + CO_2 = CaCO_3$$

$$\Delta H_R = -176.7 \text{ kJ/mol} (25^{\circ}C) \ (-42.2 \text{ kcal/mol})$$

$$= -3.15 \text{ MJ/kg CaO} \ (-750 \text{ kcal/kg CaO})$$

The acceptor, CaO, is regenerated in a separate reactor by combustion of the residual char with air, and can then be used again.

II. Performance

An experimental plant is in operation in Rapid City, South Dakota, USA. The following data are from this plant.

a. Gasifier: Coal feed: 1.13 t/h
of which 0.82 t/h is converted to gas

Gas production: 1530 m^3(iN)/h

b. Regenerator: Fuel input: 0.31 t/h
of which 0.25 t/h is burned with air

Acceptor input (+ inert material):

1. used acceptor: 5.27 t/h
2. fresh acceptor: 0.27 t/h

This yields 4.88 t/h of calcined acceptor for the gasifier

III. Feed coal requirements

a. Rank: So far lignites (Velva, N. Dakota) and subbituminous coals have been used. The coal should be as reactive as possible, and must have low contents of sulphur (FeS_2), Fe and Na.

Composition of coal

(Elementary analysis and ash content)

H	4.5 wt %
C	66.2 wt %
N	0.8 wt %
O	21.6 wt %
S	0.7 wt %
Ash	6.2 wt %

Lignite char (800°C) is removed from the gasifier and is burned with air in the regenerator as fuel. It has the following composition:

H	0.60 wt %
C	78.79 wt %
N	0.34 wt %
O	1.38 wt %
S	0.44 wt %
Ash	18.45 wt %

The suspended dust, which is deposited in the cyclone when the flue gas is freed from ash, has the following composition:

H	0.39 wt %
C	68.00 wt %
N	0.24 wt %
O	0.86 wt %
S	0.66 wt %
Ash	29.85 wt %

b. Particle size: 0.15-2.4 mm

c. Coal pretreatment: The coal is ground, dried, screened and devolatilized (coke char), so that no tar is formed in the gasifier.

IV. Raw gas composition and calorific value

Composition (mixed gas from gasification and carbonization):

H_2	55.98 vol %
CO	15.45 vol %
CO_2	10.88 vol %
CH_4	14.14 vol %
N_2	3.0 vol %
NH_3	0.12 vol %
H_2S	0.43 vol %

Calorific value: 14.65 MJ/m^3(iN) (3500 $kcal/m^3$(iN))

Flue gas from regenerator (1010°C):

H_2	0.04 vol %
CO	1.98 vol %
CO_2	28.33 vol %

N_2	68.67 vol %
H_2O	0.98 vol %
SO_2	121 ppm
H_2S	28 ppm
COS	46 ppm

V. Description

The coal is first prepared and then devolatilized. The resulting lignite char is fed into the top of the gasifier through a pressurized lockhopper; part of it is transferred to the regenerator. The acceptor material (calcined lime) passes from the regenerator to the gasifier. In the fluidized bed of the gasifier the coal meets a stream of steam and inert gas and is converted to CO, H_2 and CH_4. Any CO_2 formed reacts with the acceptor to give carbonate, while H_2S reacts to give sulphide.

The formation of carbonate liberates 176 kJ per mol, which serves to maintain the reaction temperature. Volatile hydrocarbons (35% of the C content of the coal) are decomposed to CH_4, CO, CO_2 and H_2. The residual char and the used acceptor are conveyed to the regenerator. The devolatilization char, acceptor, residual char and dolomite are conveyed by an inert gas, which consists mainly of CO_2 and N_2 and is produced by air combustion of the residual char in the regenerator (fluidized bed at 1020°C) and the simultaneous decomposition of carbonate. Continuous acceptor make-up and withdrawal is at the rate of 2% per cycle. Water-gas conversion and CO_2 removal can be omitted, since the raw gas has a high H_2/CO ratio of about 3.6. Before methanation, thorough desulphurization is necessary in order to protect the catalysts.

VI. Production of gasifying medium

Steam: Not stated

VII. Technical data

a. Carbon conversion efficiency: 94%, of which 72% corresponds to gasification and 22% to combustion

b. Gasification efficiency: 77%

c. Thermal efficiency: 92% (estimated)

d. % steam decomposition: 58%; of the remaining steam 65% can be recycled

e. Inputs for gasifier:

Steam: 4000 kg/h of live steam (500°C) and
900 kg/h of recycled steam (800°C)
= 179 kg/GJ in the raw gas
= 3.6 kg/kg coal

Acceptor: ($CaO \cdot MgO$) (+ inert material (ash, sand)
= 1/4 of the amount of acceptor)
4.88 t/h

≙ 3.15 kg/m^3(iN) raw gas

≙ 0.2 t/GJ in the raw gas

≙ 4.3 kg/kg coal

Coal: 1.13 t/h

0.05 t/GJ (in the gas)

Transferred to regenerator:	0.32 t/h residual char (at 800°C)
	5.27 t/h used acceptor ($MgO{\cdot}CaCO_3$ + CaS + CaO + inert material (800°C)
	Gas for transporting coal (400°C) 200 m^3(iN)/h
	Gas for transporting acceptor (650°C) 1440 m^3(iN)/h (flue gas)
	Flushing gas (30°C) 58 m^3(iN)/h
	Air (450°C) 2090 m^3(iN)/h
	Fresh dolomite (30°C) (+ inert material) 0.27 t/h
Removed from regenerator:	Residual ash from char combustion (1000°C) 0.07 t/h
	Calcined acceptor (1000°C) 4.88 t/h
	Flue gas (1000°C) 4180 m^3(iN)/h

The following losses occur:

Losses in suspended dust: 40 kg/h (800°C)

Losses in reject acceptor: 174.6 kg/h (800°C)

ie the following quantities of fresh dolomite are required:

0.15 kg/kg coal, or 0.11 kg/m^3(iN) gas, or 7.7 kg/GJ

f. Steam/oxygen ratio: Not applicable (no oxygen used)

g. Products:

Yield of gas: 1360 m^3(iN) raw gas/t coal

h. Pressure and temperature:	Pressure in gasifier and regenerator	11 bar
	Temperature	
	a. in gasifier	800°C
	b. in regenerator	1000°C
	Exit gas temperature	800°C
	Outlet temperature	Not stated

i. Water consumption: Not stated

j. Coal consumption for energy production: Not stated

k. Gas purification: No detailed information

VIII. Construction and development

Process development

Conoco Coal Development Co, Library, Pennsylvania

Location of pilot plant

Rapid City, South Dakota

Subcontractor for construction of plant

Stearns-Roger Corp, Denver, Colorado

Design and construction of a system for downstream gas purification and methanation

Synthetic Fuels Division, Dravo Corp,
Pittsburgh, Pennsylvania

Financial backing

Joint promotion by ERDA and American Gas Association (AGA); funding approximately $ 31 million.

The test programme is now completed; no further research is planned.

References

(1) G.P. Curran et al, Chem. Engineering Progr. 62 (2) (1962), 80

(2) C. Fink, G.P. Curran and J. Sudbury, 6th Synthetic Pipeline Gas Symposium, Chicago October 1974

(3) W.W. Bodde and K.C. Vyas, Clean fuels from coal, Symposium papers, Chicago 1973, p 70

(4) C. Fink, Clean fuels from coal, Symposium papers, IGT, Chicago 1973, p 301

(5) C. Fink, G. Curran and J. Sudbury, Clean fuels from coal, Symposium papers, IGT, Chicago 1975, p 243

(6) Office of Fossil Energy, ERDA, Quarterly Report October-December, ERDA 76 - 30 - 4

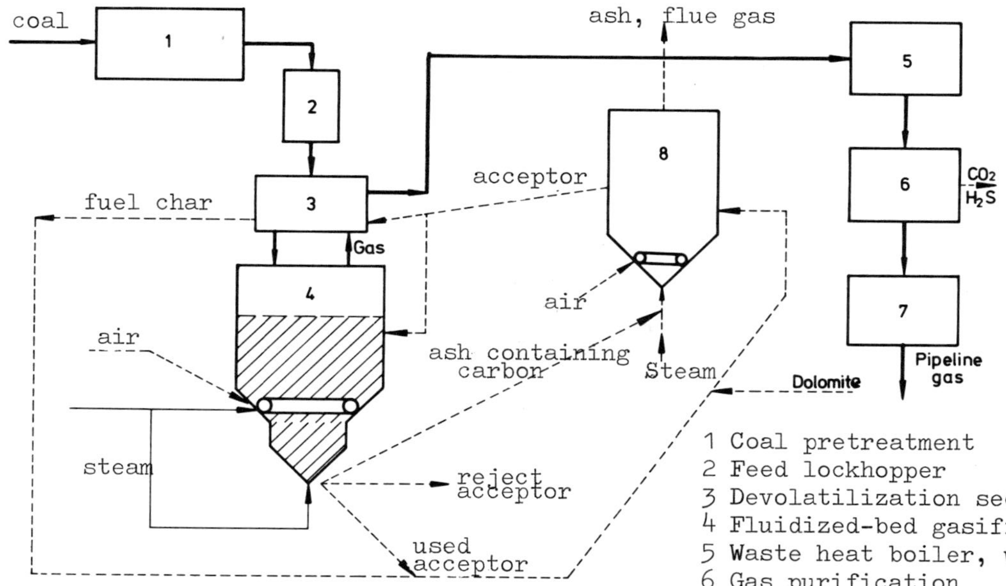

1 Coal pretreatment
2 Feed lockhopper
3 Devolatilization section
4 Fluidized-bed gasifier
5 Waste heat boiler, water wash column
6 Gas purification
7 Methanation
8 Fluidized-bed regenerator

Figure 19 Diagram of the CO_2 acceptor process

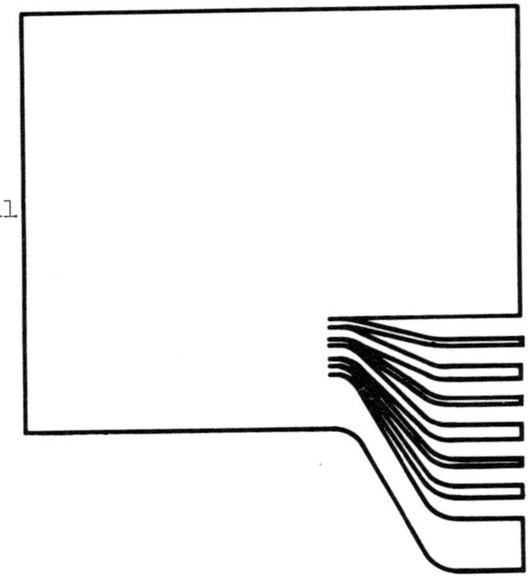

Figure 20 Heat balance for the CO_2 acceptor process

7. Bi-Gas process

I. Gasification system

Autothermic two-stage entrained-flow gasification of pulverized coal with steam and oxygen under pressure.

II. Performance

An experimental plant with a coal throughput of 4.5 t/h was completed and brought into operation at the end of 1976. It is expected to obtain about 3000 m^3 (iN)/h of SNG from this plant. The reactor has an internal diameter of about 60 cm in both gasification stages: the first stage is about 1.8 m long and the second about 8 m. The slag cooler below the gasification chambers, which is also pressurized, is about 3.7 m long. The total height is 16.3 m.

Studies have also been carried out with a semi-technical apparatus which includes only the second stage. The throughput of coal in this apparatus was about 45 kg/h.

The Foster-Wheeler Energy Corp is at present developing a combined-cycle gas/steam turbine process for producing electricity. It is planned to incorporate a gasifier corresponding to the Bi-Gas generator for this process.

III. Feed coal requirements

a. Rank: Experiments so far have used young lignites from North Dakota, older lignites from Wyoming (subbituminous C), high-volatile bituminous coal from Pennsylvania (rank A) and coals from Montana (subbituminous, Rosebud mine).

The swelling and caking properties of the coal do not affect the gasification process.

b. Particle size: 70% < 75 μm

c. Ash: The ash is discharged as molten slag. No information is given on the acceptable limits for the ash content of the coal or the ash softening point.

d. Coal pretreatment: The raw coal is pulverized, but need not be preoxidized. No mention is made of how much drying is required, if any.

IV. Raw gas composition and calorific value (dry basis)

The figures refer to results obtained with the bench-scale reactor (second stage) only. This gas is subsequently methanated to give SNG.

Composition:	CO	21.5%
	CO_2	29.3%
	H_2	32.1%
	CH_4	15.6%
	N_2	0.7%
	H_2S	0.8%

Calorific value: H_o = 13.65 MJ/m^3(iN)
(≙ 3257 kcal/m^3(iN))

V. Description

The raw coal is first pulverized and then slurried with water and graded in a cyclone. The coarse particles are repulverized wet and regraded: the fine particles are brought to the required concentration (50-60% solids in the water) in a concentrator.

The slurry is brought up to the operating pressure by piston pumps, and after preheating is introduced into the pressurized system through spray nozzles. The surface moisture on the coal dust is almost instantaneously vaporized by circulating hot inert gas. The dried coal is drawn into the gas generator by the circulating gas.

The actual gasifier comprises the two gasification stages and the slag cooler, which are separated from one another by waisted sections. The pulverized coal is blown by recirculated inert gas through nozzles into the second (upper) stage of the gasifier. Steam for gasification is also supplied through a separate ring nozzle. In the gasification chamber both of these streams are mixed with a rising stream of hot synthesis gas produced in the first (lower) stage of the reactor; the temperature of the mixture reaches about 1200°C. At this temperature the coal is devolatilized and partially hydrogasified; the residence time in the second gasification stage is about 8.5 seconds, and the temperature of the gas leaving the reactor is about 950°C. The raw gas, containing entrained residual char, is cooled to 420°C by a water spray, and the char dust is then separated out in a cyclone.

The fine char plus steam and oxygen are blown through separate tangential jets into a vortex chamber, which forms the first (lower) stage of the gasifier. In this chamber the char is rapidly and completely gasified at about 1500-1700°C, and the ash forms a molten slag. The synthesis gas produced flows through the upper opening of the chamber into the second gasification stage, where it is chilled by the fresh steam and the pulverized coal. This also serves to cool entrained drops of slag to below their melting point, and thus prevents their adhesion to the walls of the chamber.

The molten slag flows through an opening in the base of the vortex chamber into a water quencher. The opening is heated by a gas flame to prevent crust formation and to ensure a non-viscous slag. However, additional devices for breaking up the crust are also provided. Under these conditions the ash falls to the bottom of the ash cooler in small lumps and is removed periodically through a lockhopper system.

The raw gas first undergoes a second water wash in a scrubbing tower, for further cooling and dust removal. Conditions are chosen so that the gas stream retains only the moisture needed for the downstream shift conversion. Any excess of steam is condensed out at this point. In the partial conversion which follows the washing the H_2/CO ratio is adjusted to the correct value for the methanation stage. The acid gases are then removed in a Selexol unit; the hydrogen sulphide is converted to sulphur in a Claus kiln. After methanation in a fluidized bed the gas receives a final treatment for CO_2 removal in the Selexol unit, and the resulting SNG, of high calorific value, is injected into the pipeline network.

No information is given on the dynamic behaviour of the reactor.

<u>VI. Production of gasifying medium</u> Not stated

VII. Technical data

a. Carbon conversion efficiency:	Not stated
b. Gasification efficiency:	Not stated
c. Thermal efficiency:	Not stated
d. % steam decomposition:	Not stated
e. Inputs:	
Oxygen:	450 m^3(iN)/t feed coal
Steam:	1.2 t/t feed coal
Coal throughput:	4.5 t/h (for gasification only)
f. Steam/oxygen ratio:	2.68 kg/m^3(iN)
g. Products:	Not stated
h. Gasification pressure:	35-100 bar, optimum not yet known
Gasification temperature:	1500-1700°C (1st stage) 950-1200°C (2nd stage)
Exit temperature of raw gas:	950°C
i. Water consumption:	Not stated
j. Coal consumption for energy production:	Not stated
k. Gas purification:	Not stated

VIII. Construction and development

Process development (bench-scale operation, methanation)

Bituminous Coal Research Inc, Monroeville, Pennsylvania
Office of Coal Research, USA

Location of pilot plant (4.5 t/h)

Homer City, Indiana County, Pennsylvania

Project managers for pilot plant

Phillips Petroleum Co

Construction and operation of pilot plant

Stearns-Roger Inc, Denver, Colorado

Design and installation of gas generator, coal feed system and additional units

Dravo Corp, Pittsburgh, Pennsylvania

Construction of gas generator

Babcock & Wilcox Co, Fossil Power Generation Division, Barberton, Ohio

Financial backing

AGA (1/3) and DOE (formerly ERDA) (2/3); total cost $ 66 million

References

(1) R.J. Grace, Clean Fuels from Coal Symposium, Chicago Ill, 10-14 September (1973), p 179

(2) R.J. Grace, V.L. Brant and V.D. Kliewer, 5th Synthetic Pipeline Gas Symposium, Chicago Ill, 29-31 October (1973)

(3) P.B. Probert, 5th Synthetic Pipeline Gas Symposium Chicago Ill, 29-31 October (1973)

(4) Office of Fossil Energy/ERDA, Quarterly Report "Coal Gasification", October-December 1975, ERDA 76 - 30 - 4

(5) Anon, World Coal, March 1976, Vol 2, No 3, pp 31-32

(6) J.M. Miles, Oil and Gas Journal 77, 1 August (1977), 55

(7) K.E. Walker, 9th Synthetic Pipeline Gas Symposium, Chicago Ill, 31 October-2 November 1977

(8) For particulars of the Foster Wheeler combined-cycle process, see eg, Office of Fossil Energy/ ERDA, Quarterly Report "Coal Gasification", October-December 1975, ERDA 76 - 30 - 4

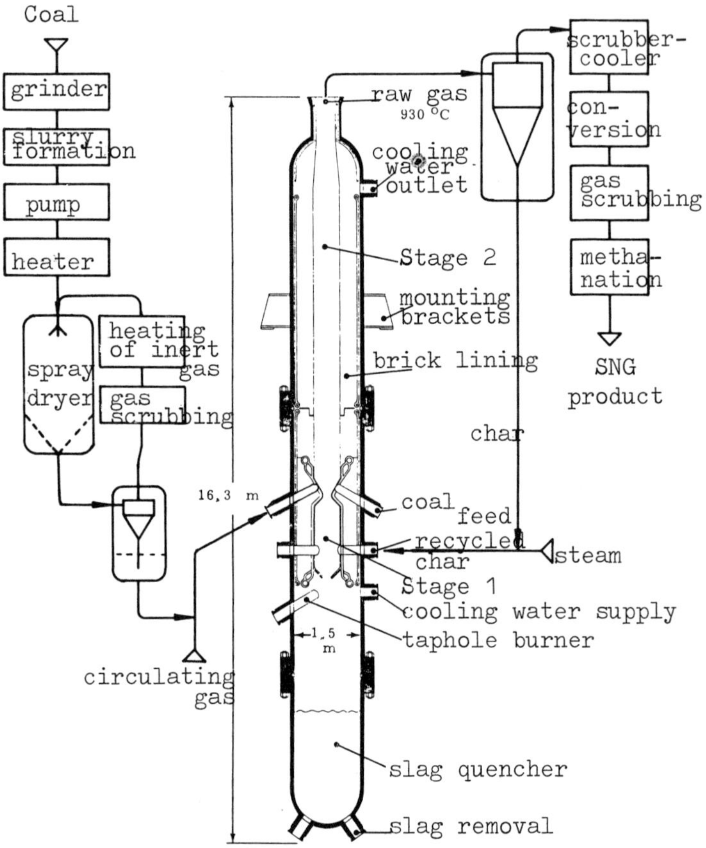

<u>Figure 21</u> Bi-Gas process (Bituminous Coal Research Inc)

8. Ash-agglomerating process

I. Gasification system

Fluidized-bed gasification of coal with steam under pressure, heat transfer being effected by pellets of ash obtained from separate combustion of coal with air.

II. Performance

An experimental plant is in operation with a throughput of 1 t/h; about 30 times this amount of heat carrier is required to convey the amount of heat needed for reaction. The plant is designed for an hourly production of 830-1250 m^3(iN) of synthesis gas.

III. Feed coal requirements

a. Rank:	The process is suitable for both caking and non-caking coals. Eastern bituminous coals are used in the experimental plant.
b. Particle size:	Char or coal < 0.15 mm is required for making the ash pellets. Coal between 0.15 and 2.35 mm is required for steam gasification.
c. Ash:	Not stated
d. Coal pretreatment:	The feed coal (< 32 mm) is dried, pulverized in a roller mill, and preoxidized in a fluidized bed at atmospheric pressure and temperatures up to 400°C.

IV. Raw gas

It is not clear to which experiments the following figures refer:

H_2	47.9-66.2 vol %
CO	0 -38.6 vol %
CO_2	3.3-28.2 vol %
CH_4	1.6- 6.9 vol %
C_2H_6	0 - 2.3 vol %
C_2H_4	0 - 1.6 vol %
$C_3+C_2H_2$	0 - 0.6 vol %
H_2S	0 - 0.2 vol %

Estimated
calorific value: up to 10 MJ/m^3(iN)
(2500 $kcal/m^3$(iN))

V. Description

The pretreated coal passes from the storage bunkers via pressurized lockhoppers into the combustion chamber or the gas generator. It is conveyed by compressed air or inert gas. In the gas generator, the coal fed in above the waisted section and the hot ash pellets introduced at about 1100°C are mixed together by steam. The coal cools these heat carriers down to about 850°C, and is partly gasified and partly devolatilized. A mixture consisting mainly of pellets with some fine char is withdrawn from the base of the generator and recirculated to the combustion chamber. There the char (plus fresh coal if required) is burned with air in a fluidized bed, producing hot pellets of heat carrier. A portion of the agglomerated ash is withdrawn continuously between the generator and the combustion chamber.

The flue gas from the combustion chamber, which is stated to be fairly dust-free, is purified in a cyclone and then expanded in a turbine. The synthesis gas is purified in a venturi scrubber, expanded and then (in the case of the experimental plant) flared off.

VI. Production of gasifying medium

a. Air: Compression and preheating

b. Steam: From the waste heat of the flue or synthesis gas

VII. Technical data

a. to e. No information is available on efficiencies, % steam decomposition or inputs.

f. Steam/oxygen ratio: Not stated

g. Products: Not stated

h. Operating pressure: 8 bar

Temperature in combustion chamber: 1100°C (softening point of the ash)

Temperature in gas generator: 850-930°C

i. Water consumption: Not stated

j. Coal consumption for energy production: Not stated

k. Gas purification: Not stated

VIII. Construction and development

Patentee: Union Carbide Corp, New York

Process development

Battelle Memorial Institute, Columbus, Ohio

Location of pilot plant (1 t/h)

Battelle's Engineering Station,
West Jefferson, Ohio

Design and construction of pilot plant

Chemical Construction Co

Financial backing

DOE (formerly ERDA), AGA

Total funding: $ 8.85 million

References

(1) W.M. Goldberger, ASME Winter Annual Meeting, Pittsburgh, Pennsylvania, November 1967

(2) W.M. Goldberger, 75th National Meeting of AIChE, Detroit, Michigan, June 1973

(3) W.C. Corder, H.R. Batchelder and W.M. Goldberger, 5th Synthetic Pipeline Gas Symposium, Chicago, 29-31 October 1973

(4) W.C. Corder and W.M. Goldberger, 6th Synthetic Pipeline Gas Symposium, Chicago, 28-30 October, 1974

(5) Anon, Hydrocarbon Proc., April 1975, p 118

(6) Office of Fossil Energy, ERDA, Quarterly Report, October-December 1975, ERDA 76 - 30 - 4

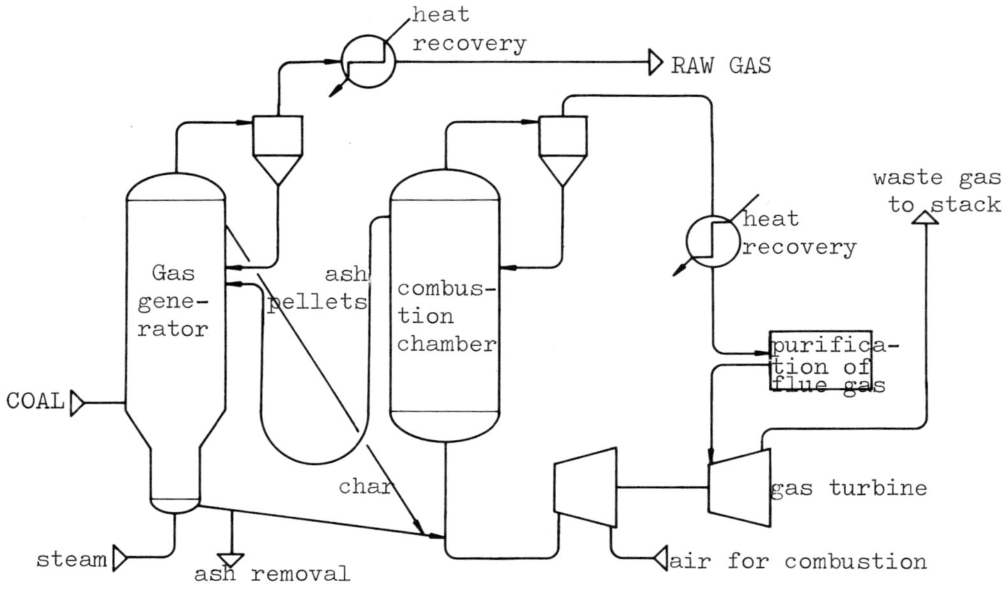

Figure 22 Ash-agglomerating process

9. Synthane process

I. Gasification system

Autothermic fluidized-bed gasification of pulverized coal with oxygen and steam under pressure.

II. Performance

Bench-scale tests have been completed. An experimental plant, designed for a coal throughput of 3 t/h and the production of 1100 m^3(iN) of SNG per hour, was brought into operation in 1976. The generator has an internal diameter of 0.9 m and a height of 21 m. The details given in Section VII and in Section V relating to the treatment of the raw gas are projected data for a plant designed to produce 6.7×10^6 m^3 (iN) of SNG per day.

III. Feed coal requirements

a. Rank:	It should be possible to use all grades of coal or lignite. The first tests were made with Montana lignite (Rosebud mine, moisture content 8-9%)
b. Particle size:	< 0.85 mm ($<$ 20 mesh). The very fine fraction ($<$ 0.075 mm) should be under 30%.
c. Ash:	Not stated
d. Coal pretreatment:	Grinding and drying

IV. Raw gas

The raw gas contains a considerable amount of methane. It is planned to process it to give SNG.

Composition (dry basis):

H_2	27.8	vol %
CO	16.7	vol %
CO_2	29.0	vol %
CH_4	24.5	vol %
C_2H_6	0.8	vol %
H_2S	0.5	vol %
N_2	0.8	vol %

Calorific value: 15.9 MJ/m^3(iN)
(3800 kcal/m^3(iN))

V. Description

Washed coal (size 0 to 19 mm) is conveyed into a bunker by inert gas through a riser. It then falls into a high-speed dryer and passes through a weigh hopper into a hammer mill, where it is pulverized to the required particle size ($<$ 0.85 mm) and dried further by inert gas at about 100°C. The circulating inert gas is separated off in a cyclone, and the coal falls from the cyclone on to a vibrating screen. It passes through a further storage bunker to one of the two pressurized lockhoppers, where it is brought up to the operating pressure of about 70 bar with CO_2. Caking coals are then preoxidized at about 400°C. Non-caking coals are transferred pneumatically by steam (70 bar) and a little oxygen directly to the top of the reactor, where they are fed in 3 m above the fluidized bed. Recently the coal has been successfully fed directly into the fluidized bed through a down pipe.

The reactor is about 21 m high, and its upper section has a diameter of 1.5 m. The upper section, about 18 m long, contains the gasification zone and the gas stabilization space, in which precarbonization of the feed coal also takes place. The height of the actual fluidized bed is up to 9 m. The gasification zone is clad with a double layer of insulation. Approximately at its centre its diameter is narrowed to 0.9 m by a diaphragm (Incoloy 800).

When the coal enters the reactor it is first precarbonized by the rising current of hot gas. The semicoke falls through the narrowed section into the fluidized bed, where it meets a stream of steam and oxygen. The gas velocity is about 0.2 m/sec. Gasification proper occurs in the fluidized bed. The residual unreacted char (ca. 35 wt %) falls from the gas producer into a char cooler, where it is cooled to about $300^{\circ}C$ by steam and, if needed, additionally by a water spray. The cooled char is discharged in a dry state. Alternatively, tests have been carried out with removal as a char/water slurry, followed by filtration. In a commercial plant the char would be used for steam production, so that the entire plant would be largely self-sufficient in steam.

The raw gas passes through a dust cyclone, a venturi scrubber, and a wash tank in which the high-boiling constituents are separated off. Further downstream are a scrubbing tower, a water-gas converter, and a hot potash scrubber, which removes most of the H_2S (down to 40 ppm) and CO_2 (down to 2 vol %).

Methanation then follows, either in a tube wall reactor (TWR) or in a hot gas recycle reactor (HGR): see reference (2).

The potash solution is treated for removal of sulphur (Stretford process) and CO_2 (50% of the latter is used again as inert gas). All liquid and gaseous waste products are burned.

VI. Production of gasifying medium

a. O_2: The small amount of oxygen required for the experimental plant is supplied by lorry transport from an oxygen plant and storage tanks.

b. Steam: The cooling of the residual char produces high-pressure steam, which is used in the process. In commercial plants steam would be produced by burning the residual char.

VII. Technical data

a. Carbon conversion efficiency:	65-80%
b. Gasification efficiency:	Not stated
c. Thermal efficiency:	Not stated
d. % steam decomposition:	Not stated
e. Inputs:	Not stated
f. Steam/oxygen ratio:	Not stated
g. Products:	Not stated
h. Operating pressure:	40-70 bar
Operating temperature:	750-1000°C
Exit gas temperature:	750°C

i. Water consumption:	In a plant designed for an SNG production of 6 700 000 m^3(iN)/day (estimated), 7000 to 11 000 m^3 of water per day would be required, which would be purified by ion exchangers.
j. Energy and coal consumption:	a. For gasification: 3 t/h b. For energy: 480 V electricity, amount not stated; also oil (for high-speed dryer, steam boiler and inert gas generator), and LNG (for inert gas generator, steam boiler and burning of wastes).
k. Gas purification:	a. Cyclone for dust removal b. Venturi scrubber c. Separation of gas and liquid in a wash tank d. Scrubbing tower with water and oil washes e. Water-gas conversion f. Scrubbing tower with hot potash solution g. Final sulphur removal

VIII. Construction and development

Process development

US Bureau of Mines, 1961

Pittsburgh Energy Research Center, Pennsylvania

Location of experimental plant (3 t/h)

Bruceton, Allegheny County, Pennsylvania

Construction of experimental plant

Rust Engineering Co, Birmingham, Alabama

Design and operation of experimental plant

The Lummus Co, Bloomfield, New Jersey

Funding

DOE (formerly ERDA): $ 6.25 million

Total cost: $ 17 million

References

(1) Th. Torkos and R. Lewis, 6th Synthetic Pipeline Gas Symposium, Chicago, Illinois, October 1974, Paper 11

(2) A.J. Forney, W.P. Haynes, J.J. Elliot, S.J. Gasior, G.E. Johnson and J.P. Strakey jr, Clean Fuels from Coal Symposium (1973), 199

(3) Anon, Engineering and Mining Journal, April 1974, 91 f

(4) D.W. Gillmore, Society of Pet. Eng. of AIME, Columbus, Ohio (1972), Paper No SPE 4158, 12

(5) H.C. Hottell, J.B. Howard, New Energy Technology - some facts and assessments, MIT, (1971), 129 f

(6) W.P. Haynes, and A.J. Forney, Clean Fuels from Coal Symposium IGT, Chicago, Illinois, 23-27 June 1975, p 149

(7) Quarterly Report of Coal Gasification, October-December 1975, ERDA Office of Fossil Energy, pp 37-41

(8) R. Lewis, E.L. Clark, W.P. Haynes and J.P. Strakey, 8th Synthetic Pipeline Gas Symposium, Chicago, Illinois, 18-20 October 1976

(9) R. Lewis, Oil and Gas Journal, 77, 1 August 1977, 31

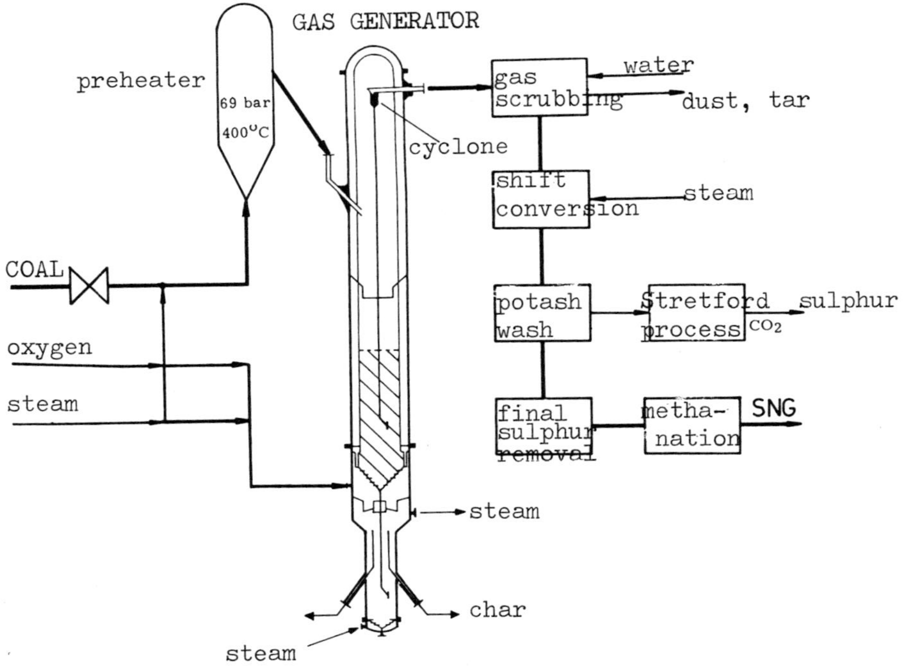

<u>Figure 23</u> Synthane process for coal gasification

10. U-gas process

I. Gasification system

Autothermic gasification of coal with steam and air or oxygen at 25 bar in a fluidized-bed reactor.

II. Performance

The process exists only on a small scale. The data are from the operation of a pilot plant with a coal input of 200 kg/h. The reactor is 7.6 m high with an internal diameter of 1.2 m. A demonstration plant is planned.

III. Feed coal requirements

a. Rank: All coals, including lignites, can be used. Tests have been carried out with Bethlehem fine coke, and COED char produced from Illinois No 6 and Western Kentucky hard coals.

b. Particle size: The raw coal is pulverized to give particle sizes < 6.5 mm. The main fraction is between 1 and 1.5 mm. The process will tolerate up to 10% of fines $< 75\,\mu$m.

c. Ash: The input coal contains on average 10 to 20 wt % ash. The ash is removed from the gasifier in the form of agglomerates between 4 and 13 mm in size, containing 90 to 95 wt % ash.

d. Coal pre-treatment: For caking coals the process incorporates preoxidation at about 400°C at the operating pressure.

IV. Raw gas

A lean gas of approximately the following composition is likely to be obtained:

Operating conditions	Steam/air	O_2/steam
H_2	9.8-15.5	34.9-36.9 vol %
CO	12.0-16.6	21.3-29.8 vol %
CO_2	10.3-14.7	34.5-42.7 vol %
CH_4	0 - 0.5	0.5- 1.1 vol %
H_2S	not stated	not stated
N_2	59.9-62.7	0
H_2O	not stated	not stated
Calorific value:		
MJ/m^3(iN)	2.83-3.64	7.44-8.53
(kcal/m^3(iN)	675-870	1778-2038)

V. Description

The predried and pulverized coal is brought up to a pressure of 20 to 25 bar in pressurized lockhoppers. Caking coals are first preheated to 400°C in a fluidized bed by heat-exchanger tubes through which steam is passed, and then preoxidized with air: non-caking coals can be fed directly into the fluidized-bed reactor, which consists of a brick-lined steel vessel.

The coal is converted to lean gas in a single-stage fluidized-bed with air and steam at about 25 bar and temperatures of around 1000°C. The conversion is rapid and the ash does not soften. Part of the gasifying medium enters the fluidized bed through a sloping base section. The remainder is fed through an opening in the centre of the base proper. High gas velocities are attained above this central point, which causes the temperature to rise locally almost to the softening point of the ash. As a result, ash-rich particles cake together, and when the weight of these agglomerates becomes sufficient to overcome the upthrust of the stream of gasifying medium they fall out of the generator into a water container. This should make it possible to obtain carbon conversion efficiencies close to those from slagging or entrained-flow gasifiers.

The raw gas produced is mixed with the gas from pre-heating and preoxidation and passes into a stabilization space above the gasification zone. The residence time is 10-15 seconds. Some steam is added to the stream of gas in order to crack tar and oils. Suspended dust is removed in a cyclone system and returned to the reactor. It is intended to attach an IGT-Meissner unit for high-temperature desulphurization, which should reduce the sulphur content to 0.23 g/m^3(iN). Hardly any details have been published for this process, which is being developed in conjunction with the Institute of Gas Technology (IGT). It is evident that Desulphurization is apparently to be effected by bringing the hot gas into contact with molten metal.

VI. Production of gasifying medium

Steam: The steam is produced in a waste heat boiler. In the preoxidation of the coal, high-pressure steam will also be obtained from the heat exchanger tubes suspended in the fluidized bed.

VII. Technical data

a. Carbon conversion efficiency: 93-98%

b. Gasification efficiency: Not stated

c. Thermal efficiency: Not stated

d. % steam decomposition: Not stated

e. Inputs:

For preheating: (At 400°C, residence time 30 min):
Air consumption: 1 kg/kg coal

For gasification: Residence time of coal: 45 min
Flow velocity: 0.3-0.75 m/s
Steam consumption: 0.5 kg/kg coal

f. Steam/oxygen ratio: Not applicable

g. Products: Not stated

h. Operating pressure: 20-25 bar, temperature: 1050°C

i. Water consumption: Not stated

j. Coal consumption for energy production: The steam obtained in the pre-oxidation of the coal is sufficient to cover the system's energy requirement.

k. Gas purification: Suspended dust is first removed in cyclones. The dust-free gas is then cooled to 450°C in a waste heat boiler, giving high-pressure steam. Next, it is intended to reduce sulphur compounds in the gas to trace quantities (< 100 ppm H_2S) by contact with molten metal. The sulphur will be recovered from the metal.

VIII. Construction and development

Institute of Gas Technology (IGT), Chicago

Location of pilot plant

Not stated; probably IGT, Chicago

Demonstration plant (design and commissioning in 1981/82)

Memphis Light, Gas and Water Division, City of Memphis (MLWG), Tennessee

Plant design:	Foster Wheeler Energy Corp
Plant operation:	Delta Refining Co
Location:	Missippi Shore, Shelby County, Tennessee

References

(1) J.W. Loeding and C.L. Tsaros, Clean Fuels from Coal Symposium, Institute of Gas Technology, Chicago, 10-14 September 1973

(2) Anon, Hydrocarbon Proc., April 1975, p 127

(3) Institute of Gas Technology (IGT) Gas Scope, Summer 1975, No 32, Chicago

(4) J.G. Patel and J.W. Loeding, Clean Fuels from Coal Symposium of the Institute of Gas Technology (IGT), Chicago, Illinois, 23-27 June 1975

(5) J.G. Patel, Oil and Gas Journal, 77, 1 August 1977, 51

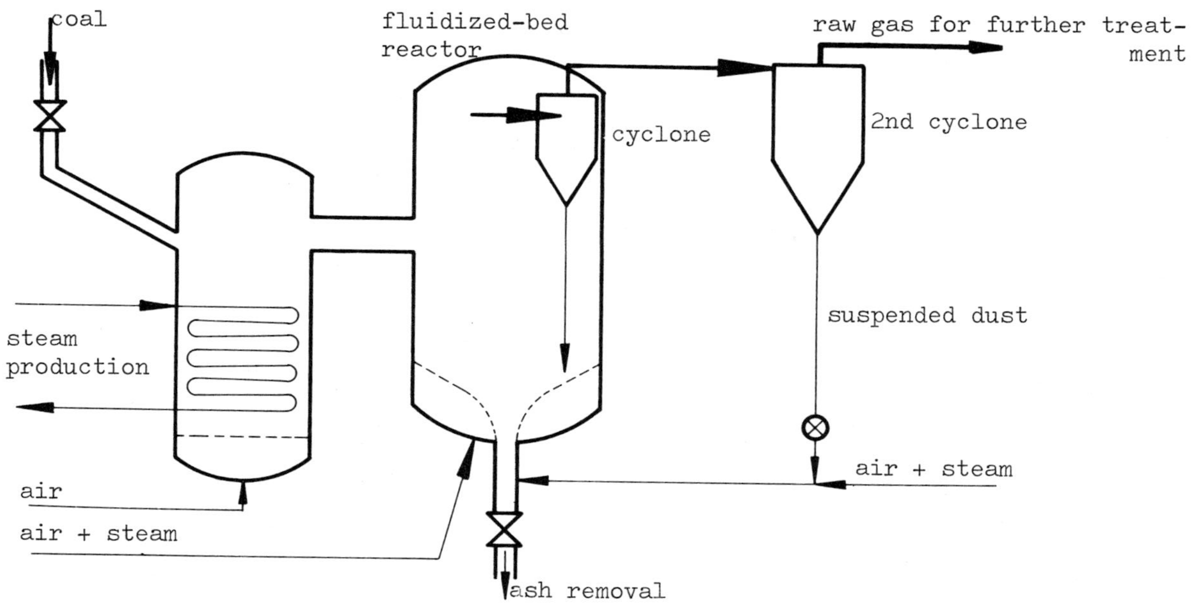

Figure 24 Simplified diagram of the IGT U-Gas process (taken from Loeding and Tsaros (1))

11. Texaco process

I. Gasification system

Autothermic co-current entrained-flow gasification of pulverized coal by steam and oxygen under pressure.

II. Performance

At the Morgantown works (West Virginia) of the Olin Mathieson Chemical Corporation, a commercial plant for the production of 6500 m^3(iN)/h of synthesis gas by the Texaco process was started up in 1956. However, no information is available about the operating data for this plant or the number of hours for which it was operational. The plant was shut down after a few years. No further large-scale plants based on coal appear to have been built.

Texaco are at present operating an experimental plant at Montebello, California, but no details have as yet been published. In the Federal Republic of Germany a large-scale experimental plant with a coal throughput of 6 t/h and a production of 12 000 m^3(iN) of raw gas per hour has been constructed at Oberhausen-Holten. This plant has now begun to operate.

The data given here and the description of the process relate to an earlier Texaco experimental plant, which was also operated at Montebello in the early 1950s.

Reactor volume:	57 l
Coal throughput:	0.27 t/h
Gas production:	620 m^3(iN)/h

III. Feed coal requirements

a. Rank: No detailed information can be found on incoming coal specifications. Data are available for tests with (A) Pennsylvania Pittsburgh seam coal, a high-volatile bituminous coal with a calorific value of 32 540 MJ/kg (7772 kcal/kg), and (B) a Japanese lignite (high-volatile C bituminous) with a calorific value of 22 840 MJ/kg (5456 kcal/kg)

b. Particle size: The coal must be finely ground (70% $<$200 mesh $\hat{=}$ 0.075 mm).

c. Ash: Coal (A) contains 7.2% ash and coal (B) 13.1%. The ash is molten: it drops from the base of the generator into a water bath and is removed continuously as granules (the melting point of the ash is not critical).

d. Coal pretreatment: Drying and pulverizing

IV. Raw gas

Composition (dry basis):

	(A)	(B)
H_2	38.7 vol %	37.9 vol %
CO	46.6 vol %	45.7 vol %
CO_2	11.5 vol %	13.2 vol %
CH_4	0.7 vol %	0.9 vol %
N_2	2.0 vol %	1.7 vol %
H_2S	0.7 vol %	0.6 vol %

Calorific value: 11.1 MJ/m^3(iN) (2650 $kcal/m^3$(iN))

V. Description

The coal is pulverized (0-74 μm) and slurried with water. The coal/water suspension (48-55 wt % coal) is brought up to the required operating pressure and heated to 540°C in a preheater. The resulting coal/steam mixture is injected into the gasification space through a nozzle at the top of the generator. Oxygen is supplied through a concentrically mounted ring nozzle. Because of the high temperature the ash melts: droplets are formed, which are cooled in a water bath at the base of the generator and removed as granules. The raw gas is taken off from the lower part of the generator: no information is obtainable on its further treatment.

VI. Production of gasifying medium

O_2:	Oxygen plant It is stated that Texaco have demonstrated the feasibility of gasification with air in the experimental plant, but no further details of this could be found.

VII. Technical data

a. Carbon conversion efficiency:	90-98%
b. Gasification efficiency:	75-80%
c. Thermal efficiency:	about 85%
d. % steam decomposition:	Not stated

e. Inputs:

	(A)	(B)	
Oxygen:	161	157	m^3(iN)/h
	590	580	m^3(iN)/t coal
	23.4	22.9	m^3(iN)/GJ raw gas

For the commercial plant, a figure of 126 m^3(iN) O_2 per Gcal in the raw gas is given (5).

Steam:	700	670	kg/t coal
	0.31	0.37	kg/m^3(iN) raw gas
	27.9	33.4	kg/GJ in the raw gas
Coal:	0.441	0.543	kg/m^3(iN) raw gas
	39.6	48.7	kg/GJ in the raw gas

f. Steam/oxygen: (A) 1.18 (B) 1.15 kg/m^3(iN)

g. Products:

Gas yield: (A) 2270 (B) 1840 m^3(iN) raw gas/t coal

Steam: Not stated

h. Pressure and temperature in the reactor: The experimental plant operated at 15 bar. In large-scale plants pressures of between 25 and 85 bar are envisaged.

Exit gas temperature: Not stated

i. Water consumption: Not stated

j. Coal consumption (experimental plant):

For gasification:	273 kg/h
For preheating:	27 kg/h
Electricity consumption:	Not stated
Surplus steam:	High-pressure steam is produced in a waste heat boiler downstream of the generator.

k. Gas purification: The raw gas is free from tar, light oils, etc, but contains dust and sulphur. The dust content is reduced to less than 1 ppm by scrubbing. Further purification employs one of the established desulphurization processes.

VIII. Construction and development

a. USA

Manufacturer

Not stated

Operators

Experimental plant:	Texaco
Large-scale industrial plant:	Olin Mathieson Chemical Corporation

Location

Experimental plant:	Montebello, California
Large-scale industrial plant:	Morgantown, West Virginia

b. Federal Republic of Germany

Manufacturers

Carl Still, Recklinghausen; F. Uhde, Dortmund

Operator

Gesellschaft für Vergasung und Verflüssigung von Steinkohle mbH
(GVV), a subsidiary of Ruhrkohle AG, Essen
Ruhrchemie AG, Oberhausen

Location

Ruhrchemie premises, Oberhausen-Holten

Funding

Total costs amount to DM 30 million; 60% of this sum is being provided by the Bundesministerium für Forschung und Technologie (BMFT) (Federal Ministry of Research and Technology) and the remainder is divided between Ruhrkohle AG (2/3) and Ruhrchemie AG (1/3).

References

(1) D. Eastman, AIME Symposium, New York, 1953, 73

(2) L.L. Newman, International Conference on complete gasification of mined coal, Liège 1951, A 1

(3) W.A. MacFarlane, J. Inst. Fuel 28 (170) (1955) 111

(4) D. Eastman, Ind. Eng. Chem. 48 (7) (1956) 1118

(5) Anon, Chem. Week 26 (1956) 76

(6) Booklet published by Ruhrkohle AG and Ruhrchemie AG, "Das RAG/RCH-Projekt Kohlestaubvergasung" (The RAG/RCH project for coal dust gasification), Essen and Oberhausen, February 1977

(7) F. Seipenbusch and P. Ruprecht, Conference on "Kohlevergasung und -hydrierung" (Gasification and hydrogenation of coal), Haus der Technik, Essen, 7-8 December 1977

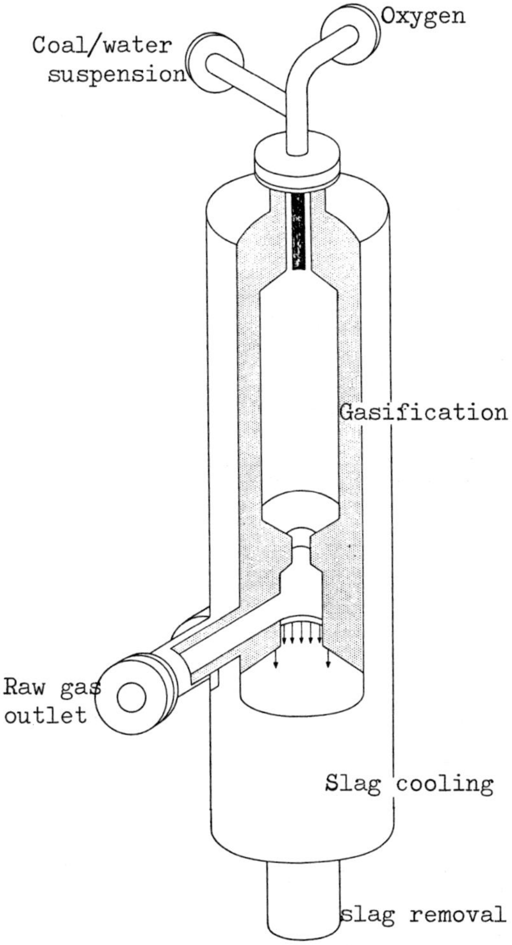

Figure 25 Simplified representation of the Texaco gas generator

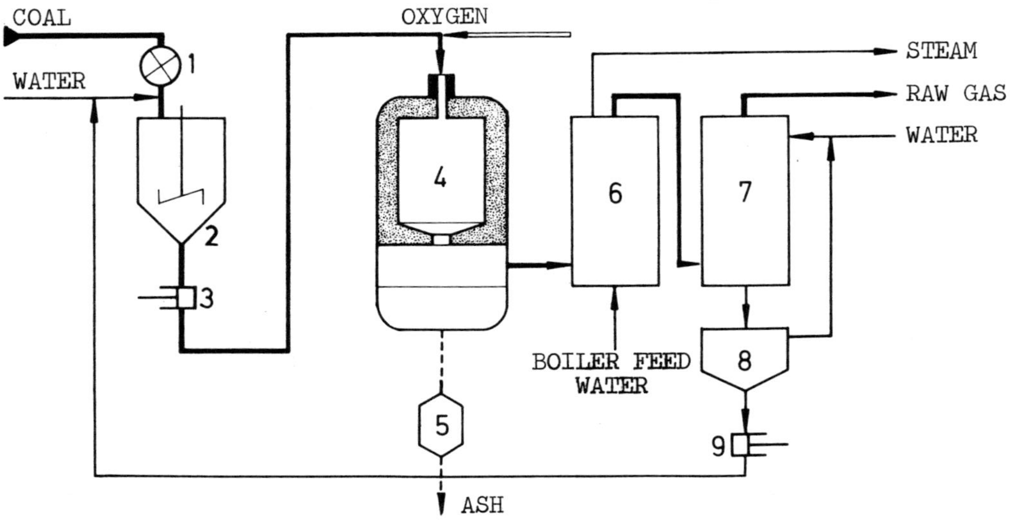

1 grinder
2 slurry container
3 slurry pump

4 gas generator
5 lockhopper
6 waste heat boiler

7 scrubber-cooler
8 concentrator
9 slurry pump

<u>Figure 26</u> Diagram of the Texaco process (experimental plant)

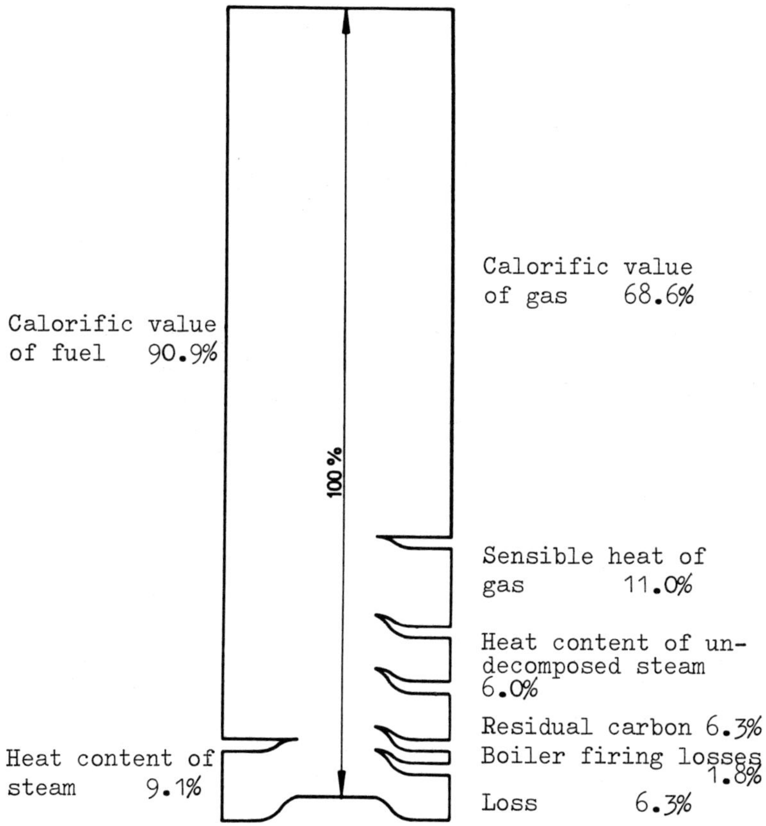

Figure 27 Heat balance for the Texaco process

12. Rummel-Otto process

I. Gasification system

Process for the high-temperature gasification of finely ground coal with air (or oxygen) and steam in a slag-bath generator at atmospheric pressure.

II. Performance

An industrial plant for the production of 16 000 m^3 (iN)/h of synthesis gas was built by Union Rheinische Braunkohlen Kraftstoff AG at Wesseling. After some difficulties, partly with the gas generator itself, a new reactor was designed and constructed at the beginning of the 1960s. This operated successfully for about 1½ years, but was closed down in 1964 in view of the general decline of interest in coal gasification. Since the autumn of 1975 a pressurized version (25-30 bar) of the same process has been under development (Saarberg-Otto process). A demonstration plant is being constructed (throughput approximately 10 t coal/h). The following data are available on the Rummel-Otto process:

a. For an experimental gasifier with an internal diameter of 0.8 m:

   Coal throughput: 2 t/h
   Gas produced: 3400-4000 m^3(iN)/h

b. Design data for the industrial plant, which had a generator of 1.8 m internal diameter:

   Coal throughput: 8 t/h
   Gas produced: 16 000 m^3(iN)/h

III. Feed coal requirements

a. Rank: In principle any type of coal with an ash content of not more than 50% can be used; lignites react more rapidly than hard coals.

Figures are available for three different modes of operation:

(A) Semitechnical tests with Rhenish lignite

(B) Semitechnical tests with Ruhr medium-rank semibituminous coal

(C) Design data for industrial operation with coal containing 15% ash

Heating value of coals (H_u):

(A)	(B)	(C)	
20.0-22.0	30.8	28.55	MJ/kg
(4800-5260)	(7360)	(6820)	(kcal/kg)

b. Particle size: 0 to 3 mm

c. Ash: The ash content should not exceed about 50%, since otherwise the slag is too viscous.

Melting point of ash: about 1250°C

d. Coal pretreatment: Preliminary drying to a residual moisture content of 6-8% is necessary.

IV. Raw gas

Composition (dry basis):

1 Tests with air
2 Tests with oxygen and steam

	(A)		(B)		(C)	
	1	2	1	2	2	Units
H_2	18.0	18.9	9.5	42.5	37.6	vol %
O_2	--	--	--	--	--	
CO	25.0	58.8	25.4	42.5	50.3	vol %
CO_2	6.5	20.3	5.0	13.6	10.4	vol %
CH_4	0.5	0.7	0.1	0.2	1.0	vol %
N_2	50.0	1.4	60.0	1.2	1.0	vol %
H_2S	not stated		not stated		0.7	vol %
Calorific values:	5.65	10.80	4.46	10.89	11.18	MJ/m^3(iN)
	(1350)	(2580)	(1056)	(2600)	(2670)	$kcal/m^3$(iN)

V. Description

The last gas generator operated at Wesseling is a reactor divided into three zones, the lower two of which are water-cooled. The first (lowest) zone is about 2 m in length and is designed as a vortex chamber. Its base is covered by a liquid-slag bath and has an overflow in the centre. The ground coal and the gasifying medium (steam and oxygen or air) are admitted to the reactor through jets built into the walls above the slag bath, which are directed tangentially downwards towards the melt. The gasifying medium is supplied through a ring main, and the ground coal is conveyed pneumatically by part of the product gas.

The flames directed on to the slag bath impart a rotary motion to the melt, which acts as a heat shield and makes it possible to use high temperatures without damaging the reactor. The method used for conveying the materials and the deflection of the flames by the slag bath ensure a high degree of turbulence which increases the reaction velocity.

During gasification part of the accumulated melt runs continuously through the overflow into a tank of water below, where it is quenched and discharged as granules.

In this first zone all the fine particles are gasified as the ash melts. The hot gas, mixed with particles of slag and char, leaves the lowest zone at 1500 to 1700°C and passes through a waisted section into the second stage. The waisted section accelerates the rotary motion of the gas, which causes drops of slag to deposit on the walls of the reactor, and from there they run down into the slag bath.

In the second zone, about 6 m long, a large proportion of the char dust entrained in the gas stream is re-gasified, causing a drop in temperature. The third zone, which follows, serves to cool and stabilize the gas. At this point circulating gas (part of the product gas) and sometimes also fine char are injected, to cool the gas and to cause entrained drops of slag to solidify. The gas leaves the reactor at 800-900°C.

The dust carried out of the reactor contains 10-30% of the solid material fed into the reactor, in the form of solidified slag and char. The greater part of it is separated out in a cyclone, which precedes waste heat recovery, and is recycled to the generator. In this way almost the whole of the mineral content of the coal ends up in the slag bath, while the carbon is almost completely gasified. The dust separator is followed by a boiler, in which the sensible heat of the raw gas is used to produce steam at about 500°C.

In the case of operation with oxygen, part of the steam is mixed with oxygen and supplied to the generator as gasifying medium. The suspended dust deposited in the steam boiler is also recycled to the generator. The gas is cooled and freed from any residual dust in a scrubber followed by a disintegrator. This also frees it from any entrained aqueous mist.

A two-shaft generator was also developed, in which allothermic gasification of the coal with steam and heating of the slag by burning residues with air are carried out separately. A generator of this kind was operated by the North Thames Gas Board.

VI. Production of gasifying medium

a. O_2: Oxygen plant (no details)

b. Steam: Steam boiler downstream of the gas generator.

VII. Technical data

	(A) 1	(A) 2	(B) 1	(B) 2	(C) 2
a. Carbon conversion efficiency:	not stated		not stated		99%
b. Gasification efficiency:	84%	94%	72%	87%	79%
c. Thermal efficiency:	not stated		not stated		83%
d. % steam decomposition:	not stated		not stated		

e. Inputs:

	(A) 1	(A) 2	(B) 1	(B) 2	(C) 2
m^3(iN)air/t coal	2000	-	3900	-	-
m^3(iN)O_2/t coal	-	330	-	540	580
m^3(iN)O_2/GJ in gas	-	16.7	-	19.1	27.0
kg steam/t coal	-	350	-	680	990
kg steam/GJ in gas	-	12.7	-	23.9	44.4
kg coal/m^3(iN) gas	0.320	0.550	0.200	0.390	0.506
t coal/GJ	0.057	0.051	0.045	0.037	0.044
f. Steam/oxygen ratio: kg/m^3(iN)	-	0.76	-	1.22	1.70
g. Products:					
Raw gas yield, m^3(iN)/t coal	3100	1830	5000	2560	2000
Steam production, kg steam/t coal:	600	700	1100	800	not stated

The steam produced has a temperature of 475°C and a pressure of 76 bar.

h. Pressure and temperature: Gasification takes place at atmospheric pressure; the temperature of the slag is 1600°C.

Exit gas temperature: 800-900°C

i. Water consumption: Not stated

j. Energy consumption: Not stated

k. Gas purification: Not stated
Removal of sulphur (H_2S) is necessary, but the gas is free from tar and hydrocarbons.

VIII. Construction and development

Rummel-Otto process

Union Rheinische Braunkohlen Kraftstoff AG, Wesseling and Dr C. Otto and Co GmbH, Bochum

Manufacturer

Dr C. Otto & Co GmbH, Bochum

Location

Wesseling, near Cologne

Saarberg-Otto process

Saarbergwerke AG, Saarbrücken, and Dr C. Otto & Co GmbH, Bochum, Saarberg/Dr C. Otto Arbeitsgemeinschaft für Kohledruckvergasung

Manufacturer

Dr C. Otto & Co GmbH, Bochum

Location

(Demonstration plant): Fürstenhausen, Saarland

Cost: DM 43 million (75% contributed by the BMFT (Federal Ministry of Research and Technology))

References

(1) R. Rummel, Ing. Techn. 28 (1956), 25

(2) R. Rummel, Coke Gas 21 (12) (1959), 492

(3) H.R. Hoy and D.M. Wilkins, British Coal Utilization Research Association Monthly Bulletin, 22 (2) (1958), 87

(4) M. MacCormac and J. Wrobel, Gas Council Research Communication G 6 113 (1964)

(5) F. Domann, Int. Z. Gaswärme 14 (1965), 325

(6) Booklet published by Dr C. Otto & Co GmbH, Bochum: Gasification of solid fuels in the Rummel/Otto Gasifier (ROG)

Saarberg-Otto process:

(1) Booklet published by Dr C. Otto GmbH, Bochum: Gasification of solid fuels in the Rummel/Otto Gasifier (ROG)

(2) M. Rossbach, P. Küffer, G. Füssmann, W. Granus and K. Peter, Conference on "Kohlevergasung und -hydrierung" (Gasification and hydrogenation of coal), Haus der Technik, Essen, 7-8 December 1977

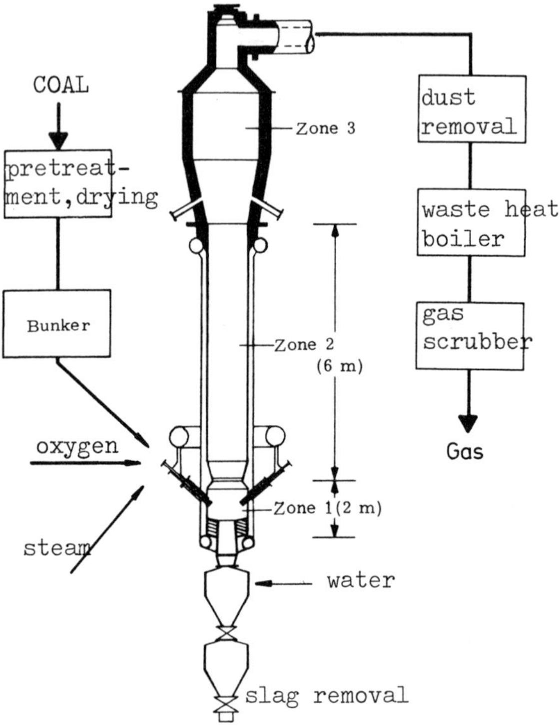

Figure 28 Diagram of the Rummel process

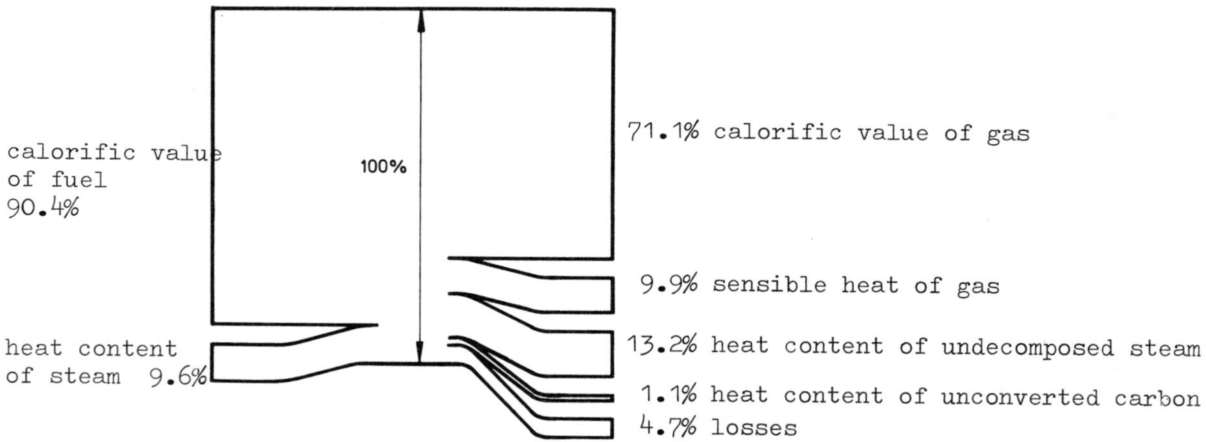

Figure 29 Heat balance for the Rummel process

13. Ruhrgas vortex chamber process

I. Gasification system

Process for the autothermic gasification of crushed coal with air and steam in an entrained-flow reactor at atmospheric pressure. The process was also tested successfully on an experimental scale for gasification with oxygen and steam to produce synthesis gas.

II. Performance

Two commercially operated plants were built for the production of heating gas. With the exception of the experimental data in Section IV b, the figures given below relate to these large-scale plants.

Up to 100 t of crushed coal per day were gasified with air in a gasifier of 2 m internal diameter and 24.5 m tall. About 30 t of suspended dust were recycled to the reactor.

Coal throughput:	4.2 t/h
Specific throughput:	1.3 t/m^2h
Gas production:	16 000 m^3(iN)/h

III. Feed coal requirements

a. Rank: In principle any coals can be used. The highest carbon conversion was obtained with lignites, though in large-scale plants the thermal efficiency was higher when coking coals were used. The data given refer to coal with the following contents of moisture and ash:

Moisture:	2.1%
Ash:	15.2%

b. Particle size: 90% smaller than 0.6 mm. The fineness of grinding has little effect. Relatively coarse dust, as in cyclone firing, can also be used.

c. Ash: The ash content is practically irrelevant, and the melting point of the ash causes difficulties only if it is very high.

d. Coal pre-treatment: Drying and grinding

IV. Raw gas

Composition (dry basis)

a. With air as gasifying medium:

H_2	8.0 vol %
CO	22.8 vol %
CO_2	5.1 vol %
N_2	64.1 vol %

Calorific value: 3.9 MJ/m^3(iN) (928 $kcal/m^3$(iN))

b. In an experimental plant, gasification of lignite with oxygen and steam gave synthesis gas of the following composition (Nistler (1)):

H_2	33.2 vol %
O_2	---
CO	48.5 vol %
CO_2	16.8 vol %
CH_4	0.5 vol %
N_2	1.0 vol %
H_2S	---

Calorific value: 10.5 MJ/m^3(iN) (2500 $kcal/m^3$(iN))

In these experiments a gasification efficiency of 75% was reached; the thermal efficiency was 87%.

In the UK, studies were also conducted (by the CEGB) on using the heating gas obtained to drive a combined-cycle gas/steam turbine process.

V. Description

Coal dust is conveyed pneumatically from the storage bunker to the gas generator. At the bottom of the gasifier (height 24.5 m, diameter 2 m), the fuel is injected tangentially through six jets, together with air preheated to 700°C by the product gas flowing counter-current. (The fresh coal is injected through four of the jets and the recycled suspended dust through the remaining two.) The slag is flung on to the water-cooled walls and drops to the bottom, where it runs out continuously into a water tank and is converted into granules; it contains 1.5-2% carbon. The raw gas flows through a waste heat boiler where it loses part of its sensible heat, and then passes through a multiclone. The waste heat boiler is U-shaped, giving a cyclone effect: it is divided into radiation and convection sections. The coal dust separated out, which may amount to as much as one third of the coal feed, is recycled to the gasifier. The gas is further purified by a disintegrator, a scrubber and a dust filter. The high temperature precludes tar formation.

VI. Production of gasifying medium

a. Air: Preheated to 700°C by raw gas flowing counter-current

b. Oxygen: Linde plant

VII. Technical data

a. Carbon conversion efficiency: Not stated

b. Gasification efficiency: 61%

c. Thermal efficiency: 84.2%

d. % steam decomposition: Not stated

e. Inputs:

Air: Not stated
(estimated) 1.6 m^3(iN)/m^3 raw gas (6100 m^3(iN)/h)

Coal: 4.16 t/h
0.26 kg/m^3(iN) of raw gas
≙ 0.067 t/GJ

f. Steam/oxygen ratio: Not applicable

g. Products:

Raw gas yield: 3740 m^3(iN) gas/t coal

Surplus steam: 2.15 t/t coal
(steam at 400°C and 35 bar, equivalent to 23% of the calorific value of the coal)

h. Operating pressure: 1 bar (atmospheric pressure)

Operating temperature: Normally above the melting point of the ash

Exit gas temperature: 700°C

i. Water consumption: Not stated

j. Other energy consumption and electricity supply: Not stated

k. Gas purification: Removal of dust and fines in the U-shaped boiler duct, multiclone, disintegrator, scrubber and dust filter.

VIII. Construction and development

Manufacturers

a. Not stated

b. Gibbons Brothers, Dudley, UK

Operators

a. Ruhrgas AG, Essen

b. National Coal Board, UK

Location

a. Recklinghausen-Suderwich, coking plant of König Ludwig mine 4/5 (dismantled 1958)

b. Wath upon Derne, Yorkshire, UK, coking plant of the Manwers Maine mine

References

(1) F. Nistler, Glückauf 88 (1952), 346

(2) K. Traenckner, Gas- und Wasserfach 93 (1952), 537

(3) Coke and Gas 19 (1957), 54

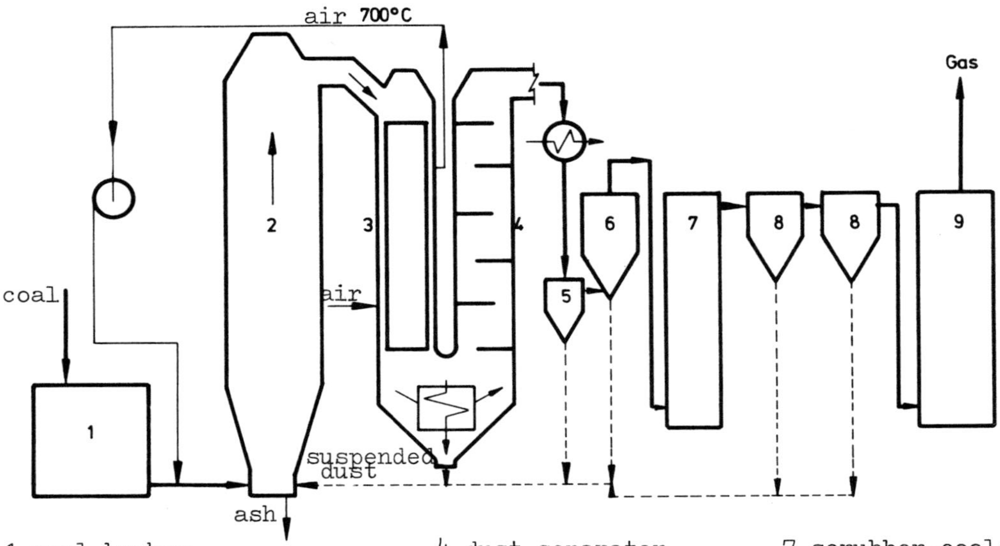

1 coal bunker
2 gas generator
3 tubular heat exchanger
4 dust separator
5 cyclone
6 multiclone
7 scrubber-cooler I
8 electrostatic precipitator
9 scrubber-cooler II

Figure 30 Ruhrgas vortex chamber process (Ruhrgas AG)

14. Combustion Engineering process

I. Gasification system

Autothermic gasification of coal with air in an entrained-flow gasifier at atmospheric pressure.

II. Performance

A pilot plant is under construction, in which the conditions for the construction and operation of an industrial plant will be determined. Construction and initial operational tests were due for completion by the middle of 1977, and an experimental programme was planned for the following year. The following data are the projected figures for the two plants. All the quantities shown are calculated or specified, and do not refer to actual operating experience.

	Pilot plant	Industrial plant	
Internal diameter (maximum)	3	9	m
Height	45	55	m
Coal input	4.5	110	t/h
Gas production	23 000	not stated	m^3(iN)/h
Energy production	78.7	" "	GJ/h

III. Feed coal requirements

a. Rank: In principle any coals can be used.

b. Particle size: The coal is pulverized and used in the form of dust. The degree of fineness required is evidently similar to that needed for cyclone firing.

c. Ash: Not stated

d. Coal pre-treatment: Drying and pulverizing

IV. Raw gas

Gasification is effected with air, but oxygen can also be used. In the former case the anticipated calorific value of the pure gas is about 4.6 MJ/m^3(iN), while with oxygen the figure is about 11 MJ/m^3(iN) (corresponding respectively to 1100 and 2670 kcal/m^3(iN)). The following is the calculated composition of the raw gas at the generator outlet when operating with air:

Composition	
H_2	10.7 vol %
O_2	--
CO	24.4 vol %
CO_2	4.1 vol %
CH_4	--
N_2	60.2 vol %
H_2S,COS	0.4 vol %
H_o	4.6 MJ/m^3(iN) (1110 kcal/m^3(iN))

It is envisaged that the purified gas obtained by gasification with air will be used for power generation. Consideration has been given to conventional under-boiler firing, to gas turbines, and to a combined-cycle gas/steam turbine process.

V. Description

The coal feed is conveyed from the storage bunker to the pretreatment unit, where it is dried and pulverized. The dry coal dust is blown into the generator, which has a maximum diameter of 3 or 9 m respectively. Its walls consist of gas-tight welded plates, which are water cooled and also have a brick lining. The reaction volume consists of a lower combustion chamber connected by a waisted section to an upper reduction zone. Fresh coal is fed tangentially to both chambers. In the combustion chamber the coal, together with recycled suspended dust recovered from the raw gas, is burned with air, which is also injected tangentially; the ash is converted to slag. The hot combustion gases stream into the upper reduction zone and provide the heat for carbonizing the fresh coal fed in there directly and also for the reduction reactions between the gas and the coal. The hot raw gas is cooled to 150°C in a heat exchanger, and then dust is removed in cyclones. The gas, now largely dust-free, is treated further in a venturi scrubber and then a Stretford hydrogen sulphide absorber. Consideration is being given to modifying the process so that the sensible heat removed from the raw gas in the heat exchanger can be transferred back to the pure gas. The ash is extracted as liquid slag from the base of the generator.

The generator of the pilot plant is identical in construction to that of the industrial gasifier described, and although the dimensions are scaled down it is still large for an experimental plant; the purpose of this is to make it possible to transpose the findings regarding flow conditions and residence time of reactants to the industrial generator.

VI. Production of gasifying medium

The air required for combustion is provided by blowers. It is preheated to about 300°C.

VII. Technical data

a. Carbon conversion efficiency:	Not stated
b. Gasification efficiency:	Not stated
c. Thermal efficiency:	Not stated
d. % steam decomposition:	Not applicable
e. Inputs:	
Coal — pilot plant:	4.5 t/h
Coal — industrial plant:	110 t/h
f. Steam/oxygen ratio:	Not applicable
g. Products:	Not stated
h. Operating pressure:	ca. 1 bar (atmospheric pressure)
Operating temperature:	
combustion chamber:	ca. 1750°C
reduction zone:	1750-900°C
Exit gas temperature:	
from generator:	900°C
from heat exchanger:	150°C
i. Water consumption:	Not stated
j. Other energy consumption and electricity supply:	Not stated
k. Gas purification:	Removal of dust and fines in cyclones and scrubbers; removal of hydrogen sulphide in a Stretford absorber.

VIII. Construction and development

Combustion Engineering Inc, Coal Gasification Fossil Power Systems, Windsor, Connecticut

Location of pilot plant

Windsor, Connecticut

Financial backing

$ 20.6 million between the end of 1974 and the end of 1979; two thirds of this are being provided by DOE (ERDA), and the remainder by Combustion Engineering Inc and by the Electric Power Research Institute (EPRI), Palo Alto, California, which is contributing $ 0.8 million.

References

(1) Low-BTU Gasification of Coal for Electricity Generation in Coal Gasification, Quarterly Report, October-December 1975, ERDA 76-30-4

(2) R.C. Patterson, The Combustion-Engineering Coal Gasification Program, Combustion 47 (1976) No 11, p 26

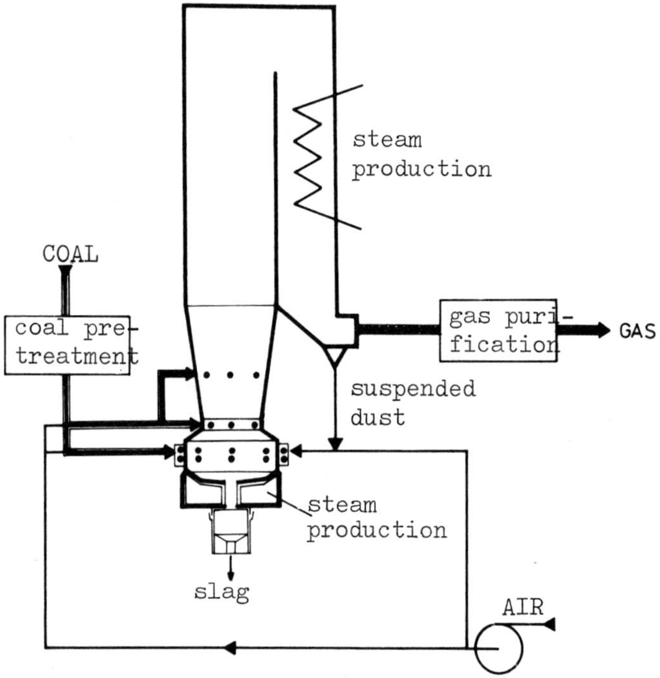

Figure 31 Combustion Engineering process

15. Westinghouse process

I. Gasification system

Autothermic counter-current gasification of char (after previous devolatilization of coal with hot raw gas) by steam and air at 10 to 20 bar in multiple fluidized beds, with simultaneous desulphurization by calcined dolomite. The process is conceived as a preliminary stage for combined-cycle gas/steam turbine processes.

II. Performance

So far only an experimental plant has been constructed.

Coal input: 600 kg/h

Gas production: 800 m^3(iN)/h

In the next stage of development a pilot plant with a coal input of 5 t/h is to be operated. The aim of the programme is to operate the combined process in a plant with a coal input of 45 t coal/h.

III. Feed coal requirements

a. Rank: Operation of the process is almost independent of coal type.

b. Particle size: 0-4 mm

c. Ash: Coals with a high content of ash can be used. The melting point of the ash is not reached.

d. Coal pretreatment: Preliminary drying (integral stage of the process)

IV. Raw gas composition and calorific value

(The figures appear to refer to bench-scale tests.)

Composition:	H_2	14.4 vol %
	CO	19.2 vol %
	CO_2	9.4 vol %
	CH_4	2.8 vol %
	N_2	54.2 vol %

Calorific value: 5.2 MJ/m^3(iN) (1250 $kcal/m^3$(iN))

V. Description

After being crushed, the coal is dried by a stream of hot gas in the first fluidized-bed unit. The dried coal is then devolatilized, desulphurized and partially gasified in a second fluidized-bed generator, also by means of hot gas. Within this fluidized bed, a circulation of char and dolomite particles is established by a riser which is mounted in the centre of the generator and fed with producer gas and fresh coal. The coal and the char are entrained by the gas stream, which has a velocity of 4.5-9 m/s. The coal is thus devolatilized. Heat is supplied to the devolatilization reactor via the hot gas. To cool the producer gas and prevent the coal from caking, the construction is so designed that the amount of char in circulation exceeds the amount of fresh coal by a factor of 100.

The producer gas flowing through the first two fluidized-bed units is obtained from the gasification and partial combustion, in a third fluidized-bed generator, of char and suspended dust from the devolatilization reactor. Air and steam serve as the gasifying and blowing agent in this third fluidized bed. The ash agglomerates and falls to the bottom, where it is removed. In addition to the transfer of heat from the gasifier to the devolatilizer via the stream of producer gas, a circulation of heated solid material can also be incorporated.

To capture and remove sulphur, the devolatilization bed is fed with calcined dolomite, which is removed continuously from the base of the generator separately and either regenerated or discarded.

VI. Production of gasifying medium

Steam:	Not stated
Air:	Compressed air will be supplied from a compressor driven by the gas turbine.

VII. Technical data

(Calculated values taken from D.H. Archer et al (3))

a.-c. Efficiencies:	Not stated
d. % steam decomposition:	Not stated
e. Inputs:	
Air (870°C)	1410 kg/h
Added steam (200°C)	340 kg/h
Coal	550 kg/h
Calcined dolomite	320 kg/h
Gas for blowing fluidized beds 1 and 2 (air + 10% N_2 at 350°C)	700 kg/h
Transport gas (for coal, char, dolomite and dust)	370 kg/h
f. Steam/oxygen ratio:	Not applicable

g. Products:

Char (from devolatilization, 870°C)	200 kg/h
Raw gas	3360 kg/h
Suspended dust	260 kg/h
Spent dolomite (870°C)	1412 kg/h
Ash	50 kg/h

h. Operating pressure: 11-17 bar

Gasification temperature:	1000-1100°C	in the gasifier bed
	700- 900°C	in the devolatilizer bed
Exit gas temperature (devolatilization):	700°C	

i. Gas purification: The dust content will be reduced to less than 50 g/GJ and the sulphur content to less than 525 g/GJ (0.23 and 2.5 g/m^3(iN) respectively). Suspended dust will be removed by cyclones.

VIII. Construction and development

Westinghouse Electric Corporation Research Laboratory, Pittsburgh, Pennsylvania

with participation by:

Public Service Indiana

Amax Coal Company

Bechtel Inc, San Francisco, California

Peabody Coal Co

Kennecott Copper Corporation

Location of the 600 kg/h plant

Waltz Mill, Pennsylvania

Planned location of the 45 t/h plant

Dresser Station of Public Service Indiana at Terre Haute, Indiana

Financial backing

Approximately $ 22.5 million, two thirds of which are being contributed by DOE (formerly ERDA) and one third by the companies participating in the development of the process.

References

(1) D.H. Archer, D.J. Vidt, D.L. Keairns, J.P. Morris and J.L. Chen, Proc. of the 3rd International Conference on Fluidized Bed Combustion, Hueston Woods, Ohio, 29 October-1 November 1972

(2) S. Lemezis and D.H. Archer, Westinghouse Engineer, July 1973

(3) D.H. Archer et al, Clean Fuels from Coal Symposium at Institute of Gas Technology, Chicago, 10-14 September 1973

(4) Office of Coal Research, Research and Development Report No 81, Interim Report No 1, 1973, Contract No 14-32-0001-1514

(5) S. Lemezis and D.H. Archer, Arch. Energiewirtsch. 28 (1974), p 651

(6) Quarterly Report Coal Gasification, October-December 1975, ERDA Office of Fossil Energy, pp 65-70

(7) R.D. Shah et al, 11th Intersociety Energy Conversion Engineering Conference, 12-17 September 1976, Stateline, Nev, Conference Papers, pp 294-299

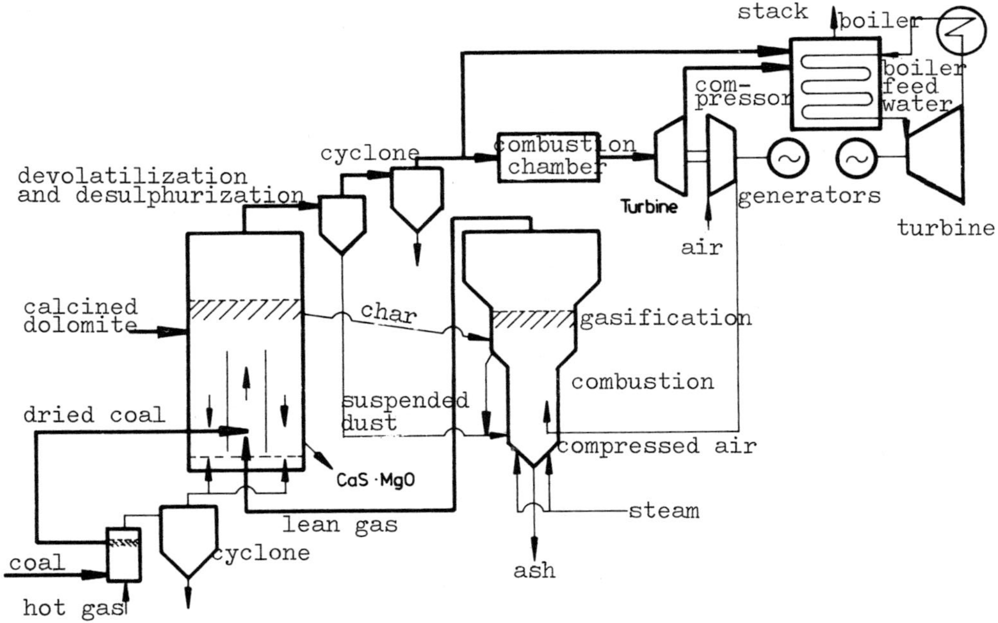

Figure 32 Westinghouse process (taken from D.H. Archer et al (3))

C. Gasification processes still at the planning stage (small-scale experiments)

1. Exxon process

I. Gasification system

Catalytic gasification of coal by steam in a fluidized bed. The necessary heat of reaction is supplied by the strongly exothermic reaction between carbon monoxide and hydrogen to give methane and steam.

II. Performance

Up to a few years ago Exxon was working on the development of a thermal coal gasification process in which coke blown to incandescence was gasified by steam. The heating of the coke was carried out separately by partial combustion with air. This development has been completely abandoned in favour of catalytic gasification.

Catalytic gasification has so far been carried out on a laboratory scale in a fixed bed and in a small fluidized-bed pilot plant. Studies have started on the design of an industrial plant.

Coal input (fluidized bed): 11 kg/h

III. Feed coal requirements

a. Rank: The process will be equally suitable for caking and for non-caking coals. Published results are for tests with Illinois No 6 coal (medium-rank bituminous), Pittsburgh seam coal (high-volatile bituminous) and Illinois coke.

b. Particle size: Not stated

c. Ash: Not stated

d. Coal pretreatment: No detailed information is given on coal pretreatment, but it is impregnated with K_2CO_3 solution before being fed into the gas generator. Reports in the literature state that tests have been carried out with different K_2CO_3 contents; impregnation with up to 20 wt % K_2CO_3 is apparently envisaged for the industrial plant.

IV. Raw gas

No information is available on gas compositions measured during the investigations. Since no air is required, the gas is nitrogen-free. The principal constituents of the raw gas are CH_4, CO_2, CO, H_2 and undecomposed steam. Small amounts of H_2S and NH_3 are also present. Since CO and H_2 are recycled, the Exxon process essentially produces only methane and carbon dioxide as gaseous products.

V. Description

The fresh coal is first impregnated with the catalyst. Just as in the preoxidation of caking coals, mixing the coal with catalyst destroys its swelling and caking properties without causing any significant decrease in its volume.

The mixture is transferred to a fluidized-bed gasifier, where it meets a mixture of steam and hot recycle gas consisting of hydrogen and carbon monoxide produced during gasification. At temperatures of between 650°C and 750°C and pressures of up to 70 bars the carbon is converted to gas.

The addition of catalyst makes it possible to carry out gasification at a lower temperature. In addition, it causes the conversion of the primary products (H_2 and CO) into methane to take place so rapidly and completely that the raw gas contains the equilibrium proportion of methane. The strongly exothermic methanation reaction then provides the heat of reaction for the endothermic steam reaction.

The raw gas, which leaves the generator at about 700°C, is first used for waste heat recovery and then processed further. Carbon dioxide, ammonia and hydrogen sulphide are removed, and carbon monoxide and hydrogen are recycled to the gasifier. Methane is the sole product gas.

The recycle gas ($CO + H_2$) is mixed with steam from the waste heat boiler and is heated to approximately 80°C above the gasification temperature. The excess temperature merely serves to cover replacement of the heat losses from the generator. The gasification reactions proper occur at about 700°C and are thermodynamically neutral overall. The rate at which the recycle gas is fed in is adjusted according to the amount of $CO + H_2$ removed with the product gas.

For economic reasons, the gasification residue, containing coal and ash, has to be treated to recover the catalyst: 60-80% of the potassium salt can be recovered by leaching with water. Leaching with acid solutions is being investigated. The recovered catalyst is returned to the process together with fresh catalyst. No plans for further utilization of the remaining residue have been reported.

VI. Production of gasifying medium

a. Steam: The steam is produced with the waste heat of the raw gas.

b. Recycle gas: The carbon monoxide and hydrogen contained in the raw gas are separated out individually and then mixed with steam and heated.

VII. Technical data

a. to c.: There are no figures available on efficiencies and steam decomposition. Since the reaction is carried out at a relatively low temperature, reduced losses and hence comparatively high gasification and thermal efficiencies are to be expected.

d. to g.: Not stated

h. Operating pressure: Up to 70 bar

Operating temperature: 650-750°C

i. to j.: Not stated

k. Gas purification: No details are available. The treatment of the raw gas does not include separate conversion and methanation stages, since these take place in the generator itself.

VIII. Construction and development

Exxon Research and Engineering Co
(Synthetic Fuels Engineering Department,
Baytown Research and Development Division)

Location

Baytown, Texas

References

(1) Anon, Engineering and Mining Journal, April 1974, 91

(2) L.E. Swabb Jr, 24th DGMK Conference (German Society of Mineralogy), Hamburg 1974
Compendium on Oil and Coal 74/75, Vol 1, p 319

(3) J.M. Eakman, T. Kalina and N.C. Nahas, Paper presented to Southwest Catalysis Society, Houston, Texas, 29 March 1976

(4) W.R. Epperly and H.M. Siegel, preprint from AIChE 11th Intersociety Energy Conversion Engineering Conference, State Line, Nevada,
12-17 September 1976

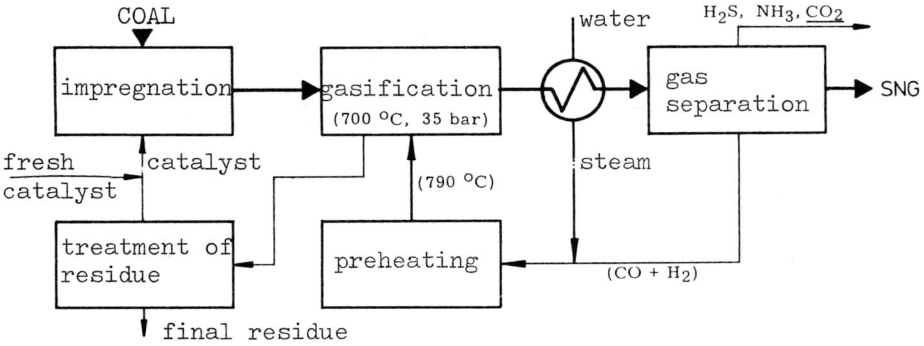

Figure 33 Exxon process for coal gasification

2. Hydrane process

I. Gasification system

Autothermic two-stage hydrogasification (dilute-phase/fluidized-bed) of unpretreated coal. The hydrogen required will be produced from the residual char.

II. Performance

The process is at the development stage. Since the beginning of 1973 laboratory experiments have been carried out with a bench-scale unit which includes the two hydrogasification stages, but not the production of hydrogen.

Reactor dimensions:	1st stage:	7.6 cm diameter, 2.16 m long (free-fall dilute-phase section)
	2nd stage:	5.1 cm diameter, 3.35 m long (fluidized bed)
Coal throughput:	4.5 to 9 kg/h	
Raw gas production:	Not stated	

III. Feed coal requirements

a. Rank:	So far tests have been carried out with Pittsburgh seam coal (high-volatile bituminous), Illinois No 6 (medium-rank bituminous) and a fairly young lignite from North Dakota. The process, however, will be suitable for all coals.
b. Particle size:	Pulverized coal is used, but no information is given on particle-size limits and distribution.

c. Ash: Not stated

d. Coal pre-treatment: Drying and pulverizing only: no preoxidation is necessary.

IV. Raw gas composition (dry basis)

H_2	18 -28	vol %
CH_4	58 -80	vol %
C_2H_6	0.1- 0.5	vol %
CO	0.5- 6.5	vol %
CO_2	0.5- 6	vol %
N_2	1.5- 2.5	vol %
H_2S	0.1- 0.4	vol %

Calorific value of raw gas: 27.2-33.9 MJ/m^3(iN)
(= 6500-8100 kcal/m^3(iN))

V. Description

The coal is injected through pressure lockhoppers into the first reactor stage, where the particles free-fall in a dilute cloud and are hydrogasified by a mixture of about 50% H_2 and 50% CH_4 flowing co-current. The residence time in this reactor is several seconds, and the temperature of the reactor wall is up to 900°C.

In this stage about 25-30% of the coal is hydrogenated, a level of conversion that makes it non-agglomerating. In the experimental reactor pressures of 70 to 150 bar have been used so far. The gas then leaves the reactor with the composition indicated above.

The porous char from this stage is transferred via a pulverizing unit to the second stage of the reactor, where it reacts with hydrogen in a fluidized bed.

The product gas from this stage contains about 50% H_2 and 50% CH_4, and is used as the gas for hydrogasification in the first stage. The gas temperature in the second stage of the reaction is about 900-1050^oC.

The residual char from the second stage is withdrawn from the reactor and gasified with steam and oxygen in order to produce the hydrogen required. Conventional processes are to be used for the hydrogen plant; no new procedure has been developed for this stage, which so far has not even been tested.

No account is given of the further treatment of the gas; it is merely stated that tar formation is very small (about 3% of the coal input) and that the tar is free from sulphur and hence well suited for firing steam boilers without pretreatment. If the product gas is to be used as SNG the remaining CO must be removed. It is proposed to do this by means of a light methanation stage, which need only be on a clean-up scale because of the relatively small amounts of CO in the gas. No information is given about the process used for sulphur removal.

When hard coals are used the residual char produced contains considerably less sulphur than the raw coal and could be used for firing without additional treatment. With lignites, however, the residual char has an increased content of sulphur, which can be attributed to the composition of the ash.

No information at all is available on the dynamic behaviour of the reactor.

VI. Production of gasifying medium

Hydrogen: The residual char is gasified with steam and oxygen.

Process steam: About half of the steam required in the process can be obtained by burning the tar. The process is presumably not self-sufficient in energy.

VII. Technical data

The only data reported are that in the experimental tests the operating pressure was between 70 and 150 bar and the added hydrogen consumption was 1.38 m^3 H_2/m^3 CH_4.

VIII. Construction and development

DOE (formerly ERDA) and US Bureau of Mines
Pittsburgh Energy Research Center, Pittsburgh, Pennsylvania

Design studies

Synthetic Fuels Division, Dravo Corp, Pittsburgh, Pennsylvania

Design and construction of a semitechnical plant

Morgantown Energy Research Center, Morgantown, West Virginia

References

(1) H.F. Feldmann and P.M. Yavorsky, 5th Synthetic Pipeline Symposium, 29-31 October 1973, Chicago, Illinois

(2) P.M. Yavorsky, Clean Fuels from Coal Symposium, 10-14 September 1973, Chicago, Illinois

(3) J.A. Gray and P.M. Yavorsky, Clean Fuels from Coal Symposium, 23-27 June 1975, Chicago, Illinois

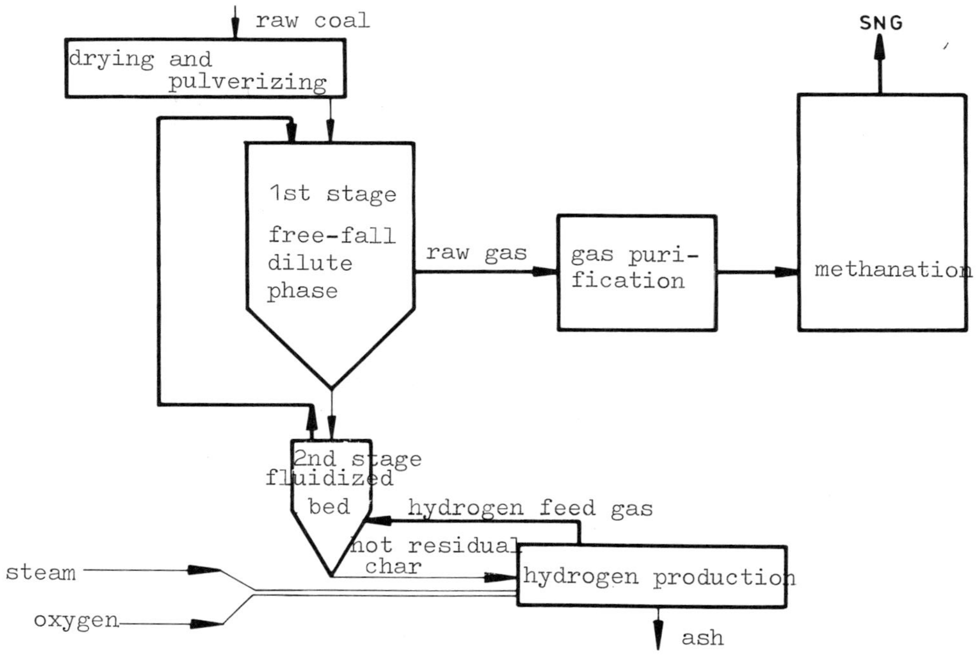

Figure 34 Hydrane process (US Bureau of Mines)

3. Molten salts process (Kellogg)

I. Gasification system

Gasification of coal with oxygen and steam in a sodium carbonate melt at high pressures. Gasification with air is also envisaged.

II. Performance

The process was first tested on a laboratory scale (internal diameter of reactor 13 cm). The next objective is a reactor with a diameter of 75 cm. The data for the small-scale unit refer to a pressure of 4 bar. The coal input was 4 t/h per m^3 of melt.

III. Feed coal requirements

a. Rank:	All coals (including caking ones) can be used.
b. Particle size:	12-20 mesh $\hat{=}$ 1.7-0.85 mm
c. Ash:	The content of ash is immaterial. The ash dissolves in the molten sodium carbonate.
d. Coal pre-treatment:	Drying and pulverizing

IV. Raw gas composition and calorific value

Composition (dry basis):

H_2	45.0	vol %
CO	33.5	vol %
CO_2	13.3	vol %
CH_4	7.5	vol %
N_2	0.4	vol %
H_2S	0.3	vol %

Calorific value: 13.0 MJ/m^3(iN) (3100 kcal/m^3(iN))

V. Description

Pulverized coal is conveyed from the coal bunker to lockhoppers, where it is put under pressure and then transported pneumatically to the generator with steam and preheated oxygen. There it is mixed with the recycle soda. Partial combustion of the coal in the molten bath provides the heat for the endothermic reaction with steam. The melt serves both as a heat transfer agent and as a catalyst. Sulphur is partly bound as Na_2S, but back reaction with CO_2 produces some H_2S. The ash at the bottom of the melt is enriched (up to 16%). Part of the melt is therefore removed continuously and processed by dissolving in water, filtering, treating with CO_2, and calcining the solid $NaHCO_3$. To protect against corrosion, the reactor is coated with Monofrax A, a type of alumina.

VI. Production of gasifying medium

Oxygen: Not stated

Steam: Not stated

VII. Technical data

a. to g.: Not stated

h. Operating pressure: 85 bar

Operating temperature: 930°C
(sodium carbonate melt)

Exit gas temperature: Not stated

i. Water consumption: Not stated

j. Coal consumption for energy production: Not stated

k. Gas purification: No tar products are formed, but sulphur must be removed before methanation.

VIII. Construction and development

M.W. Kellogg Co, Houston, Texas

Location

Not stated

References

(1) P.A. Lefrancois, K.M. Basclay and G.T. Skaperdas, Adv. in Chem. Ser. <u>69</u>, 64 Am. Chem. Society

(2) A.E. Cover, W.C. Schreiner, G.T. Skaperdas, Chem. Eng. Progr. <u>69</u> (3) (1973), 31

(3) A.E. Cover and W.C. Schreiner, Clean Fuels from Coal Symposium, 10-14 September 1973, Chicago, Illinois.

(4) L.D. Fraley and C.A. Kumar, Clean Fuels from Coal Symposium, 23-27 June 1975, Chicago, Illinois

This process is also being developed for use with air, to produce clean fuel gas for power generation, by Atomics International Division, Rockwell International Corporation, Canoga Park, California. There are plans for a pilot plant (coal input 5 t/h, operating pressure about 10 bar, gasification temperature about 1000°C) on the Norwalk Harbor power station site of the Connecticut Light & Power Company. The total cost of the project amounts to $ 6.9 million, of which 66% is being provided by DOE (formerly ERDA).

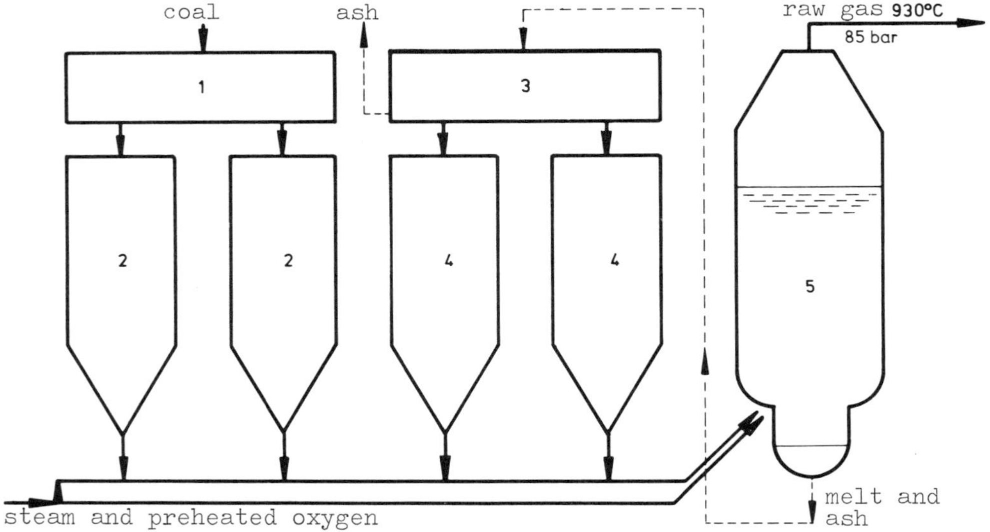

1 coal bunker
2 coal lockhoppers
3 preparation of melt
4 sodium carbonate lockhoppers
5 gas generator

Figure 35 Molten salts process (Kellogg)

4. Coal gasification in molten iron (Atgas process)

1. Two-stage gasification of coal with air for the production of lean gas

2. Patgas process with oxygen and steam as gasifying medium for the production of synthesis gas

3. Atgas process with oxygen and steam as gasifying medium for the production of SNG

I. Gasification system

Autothermic gasification of coal with air or oxygen and steam at high temperature and low pressure in a bath of molten iron and slag.

II. Performance

So far only small-scale experiments of brief duration have been carried out. These employed an induction furnace (capacity 1.8 t of molten iron) with an internal diameter of 63.5 cm. No numerical data are available.

III. Feed coal requirements

a. Rank:	This process has the advantage that it is independent of coal type. Cheap coals, which would otherwise be hardly worth treating, can be used.
b. Particle size:	The particle size should be less than 3 mm.
c. Ash:	The melting point is immaterial. The ash fraction is removed continuously as slag.

d. Coal pretreatment: Drying and pulverizing to a moisture content of 4% and particle sizes of 0 to 3 mm.

IV. Raw gas

The nature of the gas produced depends on the process used:

	Two-stage process (A)	Patgas (B)	Atgas (C)	
H_2	15	36.0	4	vol %
CO	30	63.5	--	vol %
CO_2	--	--	--	vol %
CH_4	--	--	93	vol %
N_2	55	0.5	3	vol %
Calorific values:	5.7 (1360)	12.4 (2960)	37.0 (8820)	MJ/m^3(iN) ($kcal/m^3$(iN))

V. Description

The coal is fed into the molten iron together with high-pressure steam through ceramic or oil-cooled iron lances. At 1370°C the volatile constituents are evolved explosively and are cracked in the melt. The carbon dissolves quickly in the iron and the ash passes into the slag floating on top of the iron: so does the sulphur, which forms CaS in the limestone slag. The air or oxygen which is injected a little below the surface of the molten iron reacts with carbon to give CO; the steam decomposed in the melt also reacts, forming CO and H_2.

It is reported that the gas obtained in the tests contained large amounts of dust.

The slag, which contains 4-8% of sulphur, is drawn off continuously. It would be treated with steam and processed for removal of sulphur (Claus kiln) and lime.

In the two-stage process the gas is merely freed from dust; it then heats the air for combustion in counter-current flow and is used, for example, for firing boilers.

In the Patgas process high-pressure steam is produced in the waste heat boiler. The raw gas is scrubbed, cooled, and then compressed to 70 bar for re-use.

In the Atgas process these steps are followed by conversion, thorough purification, and then methanation.

Start-up and shutdown behaviour: since the iron remains fluid for a long time because of its high heat capacity, start-up and shutdown are simplified. For the same reason, there is good flexibility in adapting to variations in load.

VI. Production of gasifying medium

1. It is planned to compress the air in two stages and then to preheat it with the raw gas.

2. The steam will be produced in a waste heat boiler.

3. The oxygen will be produced in a Linde plant with compressors driven by process steam.

VII. Technical data

The figures given are estimated values for all three processes, based on an assumed energy production of 10 500 GJ/h (2500 Gcal/h).

a. Carbon conversion efficiency: Not stated

	(A)	(B)	(C)	
b. Gasification efficiency:	98%	72%	57%	
c. Thermal efficiency:	Not stated			
d. % steam decomposition:	Not stated			
e. Inputs:	(A)	(B)	(C)	
Oxygen (4.4 bar)	--	271	373	t/h
Air	1355	--	--	t/h
Steam (4.4 bar)	--	53	70	t/h
Coal	380	450	630	t/h
Sulphur formation	14	18	25	t/h
Sulphur-free slag	46	48	63	t/h
Iron granules formed	6	15	20	t/h
Added limestone	24	32	42	t/h
f. Steam/oxygen ratio:	--	0.28	0.27	kg/m^3(iN)
g. Products:	Not stated			
h. Operating pressure:	4.4 bar			
Operating temperature:	1400°C			
Exit gas temperature:	1100°C			

i. Water consumption: Not stated

j. Coal consumption for energy production: Not stated

k. Gas purification: Since the gas is almost sulphur-free (50 ppm SO_2), no desulphurization is necessary. Trace elements present in the coal (Ni, V) are dissolved in the molten iron or in the slag.

The dust content of the gas appears to be quite high; 95-97% will be removed by electrostatic filters.

VIII. Construction and development

Applied Technology Corporation, Pittsburgh, Pennsylvania

Financial backing

American Gas Association (AGA)

ERDA (formerly Office of Coal Research)

Environmental Protection Agency (EPA)

Development is not being funded further.

References

(1) P.J. la Rosa and R.J. McGarvey, Clean Fuels from Coal Symposium Papers, September 1973, p 285

(2) Office of Coal Research, 1973 Annual Report, Clean Energy from Coal, p 19

(3) J.A. Karnavas, P.J. la Rosa and E.A. Pelczarski, AGA 1972, Pipeline Gas Symposium, October 1973

(4) "A Survey of R & D Projects directed towards the conversion of coal to gaseous and liquid fuels" prepared for AGA by the Institute of Gas Technology, p I - A - 25, November 1972

Postscript on coal gasification in molten iron

Experiments carried out by Eisenwerk-Gesellschaft Maximilianshütte mbH in Sulzbach-Rosenberg indicate that gasification in molten iron poses far fewer problems if the Oxygen-Bodenblasen-Maxhütte (OBM) process is used (bottom-blown). In this modification the immersed lances for introducing coal and oxygen, which are particularly liable to give trouble, are replaced by injection tubes situated in the base of the crucible. Since the OBM process is used in the manufacture of steel in bottom-blown converters, a substantial body of experience is already available for the development of coal gasifiers of this type. The first brief tests by Maxhütte (duration about 30 minutes, coal input 2.5 t) in a 5 t converter showed good results. The composition of the raw gas is approximately that given for the Patgas process, and desulphurization is also good (50-100 ppm H_2S in the raw gas). If the coal contains not more than 7% residual moisture the process requires no energy supplement. No steam is needed. However, if the coal has a high dross content it is apparently difficult to control the heat balance in the iron-bath reactor. Since 250 t converters are already in use in the steel industry it should also be possible to operate very large coal gasifiers (200 000-400 000 m^3 (iN) raw gas/h).

References

(1) Private communication

(2) DOS 25 20 883

(3) Gas- und Wasserfach - Gas/Erdgas 118 (1977), 350

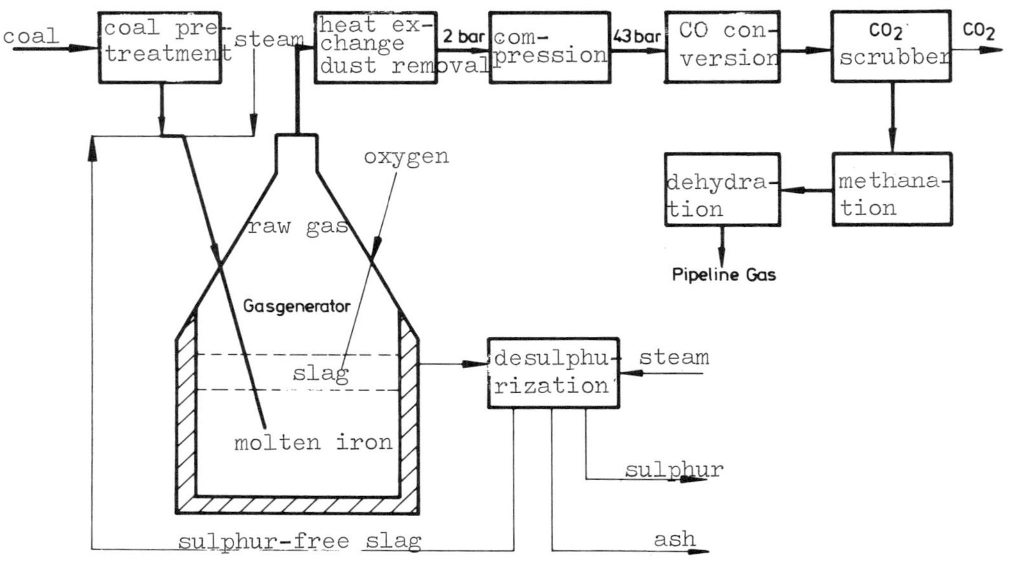

Figure 36 Atgas process

D. Gasification of coal using nuclear process heat

In all the processes described so far for coal gasification about 30% of the input coal feed has to be burned, directly or indirectly, in order to obtain the heat required for the gasification process. In the light of the high price of coal and the enormous advances made recently in nuclear engineering, the possibility of using heat from high-temperature nuclear reactors for gasification processes was discussed as early as the 1960s.

The advantages of this technology can be summarized as follows:

a. All the coal is used for the gasification process itself.

b. The use of expensive and energy-intensive oxygen is eliminated.

c. Because of the reduced coal consumption, atmospheric pollution associated with the product gas is reduced by about 30%.

Development work on these techniques is being carried out jointly, mainly with financial backing from the Bundesminister für Forschung und Technologie (Federal Minister for Research and Technology), by the Nuclear Research Centre at Jülich, Rheinische Braunkohlenwerke AG in Cologne and Bergbau-Forschung GmbH in Essen. The Nuclear Research Centre at Jülich is aiming to develop a high-temperature reactor which will supply process heat at as high a temperature as possible, Rheinische Braunkohlenwerke AG is investigating the hydrogasification of lignites and hard coals, and Bergbau-Forschung is working on steam gasification. Experiments are at present being carried out on a semi-technical scale, and no operational data for large-scale plants are available as yet.

The information given here is taken from a comprehensive study project for the Landesminister für Wirtschaft, Mittelstand und Verkehr des Landes NRW (Nordrhein-Westfalen Regional Minister for Economic Affairs, Small Businesses and Trade). This study gives conceptual designs for large-scale plants for coal gasification using nuclear process heat, based on laboratory data. However, since the details are not yet settled and may well be modified in the light of subsequent developments, we have given only highly simplified diagrams of the processes involved, designed to show their principal features.

1. Steam gasification of lignite

I. Gasification system

Allothermic gasification of previously carbonized lignites in a fluidized-bed reactor with a mixture of carbonization gas and steam at 40 bar, using nuclear process heat.

II. Performance

The process was designed as part of a plant for the direct reduction of iron ores in a study carried out for the Nordrhein-Westfalen Regional Ministry of Economic Affairs, Small Businesses and Trade. It is intended to produce about 136 t of iron sponge per hour from enriched iron ore pellets. For this purpose the gasifier would have a throughput of 150 t dry lignite/h and a gas production of about 230 000 m^3(iN)/h of dry raw gas.

The data are, however, essentially derived from calculations based on thermodynamics and heat technology. Some experiments carried out in the laboratory suggest that the design data reported for the plant will require further modification. The figures given below are based on the data given in the report.

III. Feed coal requirements

a. Rank: The design data are based on a Rhenish dry lignite with a residual moisture content of 18% (TBK 18).

Analysis:	C	52.6 wt %
	H	4.0 wt %
	O	19.4 wt %
	N	0.8 wt %
	S	0.4 wt %
	Ash	4.8 wt %
	H_2O	18.0 wt %

b. Particle size:	Not stated
c. Ash:	The design data refer to the coal described above. No information is given on other coals, or on the permissible content and melting behaviour of the ash.
d. Coal pre-treatment:	Not stated

IV. Raw gas composition (dry, tar-free basis)

There were considerable discrepancies between the raw gas composition as calculated thermodynamically and that found in non-isothermal experiments:

	calculated	from experiments
H_2	37.9 vol %	67.1 vol %
CO	5.4 vol %	3.0 vol %
CO_2	33.6 vol %	28.3 vol %
CH_4	23.1 vol %	1.6 vol %
Calorific value:	14.7 MJ/m^3(iN) (= 3508 $kcal/m^3$(iN)	9.58 MJ/m^3(iN) = 2288 $kcal/m^3$(iN))

The calculated values were used for the plant design. The carbonization gas has the following predicted composition (dry, tar-free basis):

CO	53.8 vol %
CO_2	32.2 vol %
CH_4	9.0 vol %
N_2	4.1 vol %
H_2S	0.9 vol %
Calorific value:	10.6 MJ/m^3(iN) (= 2523 $kcal/m^3$(iN)

V. Description

The finely ground lignite is conveyed pneumatically from the storage bunker to the pressurized lockhoppers. From there it passes into a fluidized-bed carbonizer, through which flows steam (43 bar/535°C) from the steam boiler.

Heat transfer is effected by a heat exchanger immersed in the fluidized bed. The helium flowing through the heat exchanger has been heated by heat from the nuclear reactor, and has already passed through the gas generator and intermediate superheater before entering the carbonizer.

The helium enters the carbonizer at 742°C and is cooled down to 704°C. It reaches the steam boiler at the latter temperature. The feed coal is expected to lose 46.8% of its weight in the carbonizer.

The hot, devolatilized char drops through an overflow duct into the gasifying generator. The carbonization gas and the steam which it contains are heated from the carbonization temperature of 400°C to the gasification temperature of 660°C in an intermediate superheater. They then serve as the fluidizing medium for the char in the gasifier, which is likewise constructed as a fluidized-bed reactor. There is an immersed heat exchanger in this fluidized bed also, through which helium is circulated to supply heat for the endothermic steam gasification reaction. The helium (43 bar) enters the gasifier at 900°C and leaves it at 796°C.

The ash, which still contains about 50 wt % C, is discharged through a lockhopper system of the conventional type.

The hot raw gas is first freed from coarse dust in a cyclone unit. It then passes through a waste heat boiler which produces high-pressure steam, and enters (at 250°C) a tar scrubber, in which high-boiling tars and fines are washed out above the steam dew-point. The raw gas is then cooled to 80°C in further waste heat coolers, which serve to produce low-pressure steam and to pre-heat boiler feed water. The low-boiling tars and condensates which separate out in these coolers are largely dust-free and therefore run cleanly off the cooling surfaces.

The product gas is further purified by a Rectisol wash. For this purpose it is first cooled to -65°C, and the hydrogen sulphide and part of the carbon dioxide are dissolved out with methanol in the first main scrubbing column. Most of the remaining CO_2 is extracted with freshly regenerated methanol in a second column, and the last traces are removed by a molecular filter. The used methanol from both scrubbing stages is regenerated again by heating followed by flashing, and the CO_2 obtained in the second scrubbing stage of washing is used to seal the coal and ash ports lockhoppers. The acid gas from the first scrubbing stage of washing, which is rich in H_2S, is treated for elementary sulphur recovery in a Claus kiln.

The purified gas is finally separated, in a low-temperature fractionation plant, into a hydrogen fraction, which constitutes the reducing gas required, and a second fraction containing mainly carbon monoxide, nitrogen, dissolved hydrogen and various hydrocarbons, which is used as a fuel gas for heating the reducing gas: the third fraction contains the high-calorific components such as methane and other hydrocarbons, which after further treatment are injected as SNG into the public gas mains.

VI. Production of gasifying medium

Steam:	The high-pressure steam required for the carbonizer is obtained from the steam generator downstream of the gasifier. After superheating, the mixture of steam and carbonization gas is used as the gasifying medium.

VII. Technical data (gas production only)

a. Carbon conversion efficiency: 96.7%

b. Gasification efficiency: Not stated

c. Thermal efficiency (including gas treatment): 84.3%

d. % steam decomposition: 28.9%

e. Inputs:

Dry lignite (TBK 18):	0.652 kg/m^3(iN) dry raw gas 44.4 kg/GJ in raw gas
High-pressure steam: (43 bar/535°C)	2.00 kg/kg TBK 18 1.30 kg/m^3(iN) dry raw gas 88.9 kg/GJ in raw gas

f. Steam/oxygen ratio: Not applicable (allothermic process)

g. Products:

Gas yield:	1533 m^3(iN) dry raw gas/t TBK 18
Product gases (separated and purified):	492 m^3(iN) reducing gas (H_2)/t TBK 18
	332 m^3(iN) SNG/t TBK 18
	3.05 GJ fuel gas/t TBK 18
Tar, oil, etc:	Not stated
Waste heat (low-pressure steam):	2.81 GJ/t TBK 18
Waste heat (pre-heating of boiler feed water):	0.96 GJ/t TBK 18

h. Gasification pressure: 40 bar
Gasification temperature: 660°C
Exit gas temperature: 660°C
State of gasifying medium: High-pressure steam mixed with carbonization gas; 660°C, 40 bar

i. Water consumption: The plant has to be supplied with 300 t/h of high-pressure steam for gasification. A further 740 t/h of water are required in the waste heat system, but are recovered as steam of various grades. 229 t/h of the gasification steam appear as gas liquor. It is not stated whether direct recirculation after suitable treatment is possible.

j. Other energy consumption:

Process heat from high-temperature nuclear reactor: 3.29 GJ/t TBK 18

	2.14 MJ/m^3(iN) dry raw gas 146 MJ/GJ in raw gas
Electricity:	20 MW 133 kWh/t TBK 18 87 Wh/m^3(iN) dry raw gas 5.9 kWh/GJ in raw gas
k. Gas purification:	Dust removal is carried out above the steam dew-point. The coarse dust is separated in a cyclone unit immediately following the gasifier, and fines are removed by a tar scrubber. Low-boiling tars and water vapour condense out in the cooling stages which follow the scrubber, and run freely off the cooling surfaces. The gas is then purified further by several stages of washing with methanol at temperatures down to -70°C under a pressure of 40 bar (Rectisol process). Impurities such as crude benzene, hydrogen sulphide, carbon dioxide and water vapour are taken up by the methanol, which is then regenerated by heating and flashing. The impurities in the waste gas thus produced are recovered separately. After purification the gas is finally fractionated at low temperature into hydrogen, a methane-rich fraction as a basis for SNG, and a fuel gas consisting mainly of carbon monoxide with some hydrogen, nitrogen and methane.

VIII. Construction and development

The above process design formed part of a study on the use of high-temperature nuclear process heat.

References

(1) Kohlevergasung mit nuklearer Prozesswärme (Coal gasification using nuclear process heat), Düsseldorf, 1974:
A study carried out for the Nordrhein-Westfalen Regional Minister for Economic Affairs, Small Businesses and Trade by a consortium formed by Gesellschaft für Hochtemperaturreaktor-Technik mbH, Bensberg; Bergbau-Forschung GmbH, Essen; Rheinische Braunkohlenwerke AG, Cologne; and Steag AG, Essen.

(2) Prototypanlage nukleare Prozesswärme (PNP) (Prototype plant for using nuclear process heat), December 1976:
Progress report at the end of the design phase, prepared by the consortium formed by Bergbau-Forschung GmbH, Essen; Gesellschaft für Hochtemperaturreaktor-Technik mbH, Bensberg; Hochtemperatur-Reaktorbau GmbH, Mannheim; Kernforschungsanlage Jülich GmbH, Jülich; and Rheinische Braunkohlenwerke AG, Cologne.

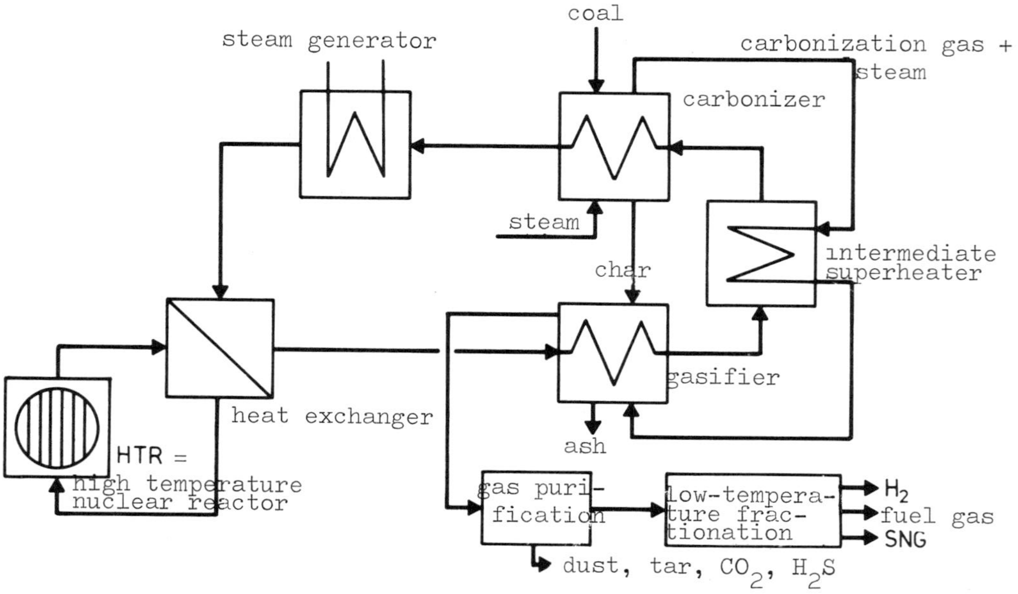

Figure 37   Flow-sheet for the steam gasification of lignite

2. Steam gasification of hard coal

I. Gasification system

Allothermic gasification of hard coal in a fluidized-bed reactor with steam at 40 bar, using nuclear process heat. The gasification reactor is preceded by a carbonizer, which is also operated with heat from the high-temperature nuclear reactor.

II. Performance

Bergbau-Forschung GmbH in Essen is operating a small-scale gasification reactor with a char throughput of up to 5 kg/h, and also a semi-technical plant for a throughput of 100-200 kg/h. No results are available for the latter at present. Operating experience with the small-scale reactor and results of laboratory experiments form the basis of a study being carried out for the Nordrhein-Westfalen Regional Ministry of Economic Affairs, Small Businesses and Trade with the object of designing a large-scale industrial plant with a throughput of 250 t coal/h and an hourly production of 155 000 m^3(iN) of SNG together with 47 800 kg of tar and oil products. The following data refer to this plant.

III. Feed coal requirements

a. Rank: The study is based on a high-volatile bituminous coal ("Hagen") of the following composition:

Pure coal	84 wt %
Water	10 wt %
Ash	6 wt %
Volatile constituents (dry basis)	36 wt %

Analysis of the pure coal (dry basis)

C	85.5 wt %
H	5.0 wt %
O	7.0 wt %
N	1.5 wt %
S	1.0 wt %
Calorific value (raw coal)	28.9 MJ/kg (= 6900 kcal/kg)

b. Particle size: In the large-scale plant the coal will be pulverized to a particle size of < 0.1 mm.

c. Ash: The study refers only to the above coal containing 6% of ash. No information is included on the maximum acceptable ash content of the coal, or the effect of the composition of the ash on the reactivity of the coal or the conditions for operating the reactor.

d. Coal pretreatment: The coal named in the report has to be pulverized. Preliminary drying may also be necessary for other coals.

IV. Raw gas composition

The feed coal is first carbonized, giving a carbonization gas of the following composition (dry, tar-free basis):

H_2	24.2 vol %
CO	14.5 vol %
CO_2	1.9 vol %
CH_4	44.2 vol %
higher hydrocarbons	13.3 vol %
H_2S	1.2 vol %
NH_3	0.7 vol %

The carbonization gas also contains 0.802 kg of tar and oil per m^3(iN) of dry gas of the above composition.

The clear gas obtained from gasification of the devolatilized char has the following composition (dry basis):

H_2	57.5 vol %
CO	19.8 vol %
CO_2	20.0 vol %
CH_4	1.9 vol %
H_2S	0.14 vol %
NH_3	0.69 vol %
Calorific value:	10.6 MJ/m^3(iN) (= 2537 kcal/m^3(iN))

After tar, oil, water and ammonia have been condensed out, the carbonization gas is combined with the clear gas from the gasification stage before being further purified by methanol scrubbing. The mixed gas thus obtained has the following composition:

H_2	54.4 vol %
CO	19.4 vol %
CO_2	18.2 vol %
CH_4	6.3 vol %
higher hydrocarbons	1.4 vol %
H_2S	0.3 vol %
Calorific value:	12.9 MJ/m^3(iN) (= 3090 kcal/m^3(iN))

V. Description

After being pulverized the raw coal is conveyed pneumatically to the storage bunkers and then on to the coal hoppers, from which it is fed into the carbonizer. The carbonizer* is operated as a fluidized bed at a pressure of 40 bar. The fluidized bed is maintained by the circulation of carbonization gas. An amount of gas equivalent to that freshly produced is withdrawn continuously from circulation and freed from dust in a high-temperature scrubber. The scrubbing liquid used is tar that is introduced into the circuit. The gas leaves the tar scrubber at about 250°C and is cooled further to 70°C in order to separate out low boiling tars and oils.

The devolatilized char leaves the carbonizer by an overflow duct and falls into the gasifier, which is likewise constructed as a fluidized-bed reactor, where it meets a stream of superheated steam (790°C, 43 bar). The gas produced here, which is free from tar and higher hydrocarbons, undergoes dry dust removal in a cyclone, and after passing through a waste heat boiler and a gas preheater for the methanation stage it is combined with the carbonization gas.

The ash is discharged from the gasifier in a dry state through a system of lockhoppers. The mixed gas consisting of the carbonization and clear gases is subjected to several stages of scrubbing with methanol under pressure (Rectisol wash). The first stage removes residual moisture and traces of higher hydrocarbons, while two subsequent stages free it from the

---

*In Figure 39 the carbonizer and the gasification reactor have been combined for the sake of simplicity, and are referred to as the gas generator.

acid gases CO_2 and H_2S; the latter is then converted to elementary sulphur in an integral Claus kiln. The pure gas is converted to SNG in a methanation unit at 400°C, and after final cooling and condensing out of the water produced in the reaction, it is fed into the mains network. The treatment of the gas produces large quantities of high-pressure steam and hot water, which are supplied to an adjoining steam power station.

VI. Production of gasifying medium

Steam: The steam required for gasifying the devolatilized char (43 bar, 790°C) is obtained from waste heat boilers and conditioned in a superheater heated by helium.

Circulating carbonization gas flows through the carbonizer: an amount of gas equivalent to that freshly produced is withdrawn continuously.

VII. Technical data

a. Carbon conversion efficiency: Not stated

b. Gasification efficiency: Not stated

c. Thermal efficiency: Not stated

d. % steam decomposition: 45%

e. Inputs:

Raw coal: 0.428 kg raw coal/m^3(iN) dry mixed gas*
33.1 kg raw coal/GJ in dry mixed gas

---

*The calorific value of the mixed gas does not take tar and oil by-products into account.

Steam: (43 bar/790°C)	2.17 t/t raw coal 0.927 kg/m^3(iN) dry mixed gas 71.7 kg/GJ in dry mixed gas
f. Steam/oxygen ratio:	Not applicable (allothermic process)
g. Products:	
Gas yield	2338 m^3(iN) dry mixed gas/t raw coal
Tar and oil	195 kg/t raw coal 83.2 g/m^3(iN) dry mixed gas
Ammonia	10.7 kg/t raw coal 4.6 g/m^3 (iN) dry mixed gas
High-pressure steam* (43 bar/530°C)	0.992 t/t raw coal 3.48 GJ/t raw coal
Hot water* (45 bar/180°C)	4.99 t/t raw coal 3.82 GJ/t raw coal

h. Gasification pressure: 40 bar

Gasification temperature:	790°C
State of gasifying medium:	Steam at 43 bar, 790°C Carbonization by circulating gas at 40 bar
i. Water consumption: (including treatment of gas to produce SNG)	1 m^3/t raw coal, assuming extensive treatment and re-use of effluent

j. Other energy consumption:

Process heat from high-temperature nuclear reactor:	8.16 GJ/t raw coal
Heat content of fresh water condensed in steam turbines:	0.86 GJ/t raw coal
Electricity consumption:	477 kWh/t raw coal

---

*These figures include complete purification of the gas and its further treatment to give SNG.

k. Gas purification: Since it is intended to obtain tars and oils as by-products in this plant, the carbonization gas and the gas produced by gasification (clear gas) undergo dust removal separately and are not combined until they enter the unit for further purification (Rectisol wash).

The gas obtained from carbonization is first crudely freed from dust by scrubbing. It is then cooled, when tar, oil and gas liquor separate out and are drawn off. After this it is purified further together with the clear gas.

The product of gasification (clear gas) undergoes dry dust removal in cyclones; it then passes through the waste heat boiler and is combined with the carbonization gas. The mixed gas is cooled to $5^{o}C$, when the remainder of the gas liquor condenses out, and then undergoes several stages of further purification in a Rectisol plant. The first stage removes residual water and any hydrocarbons present. The second stage removes H_2S, and the last stage any remaining CO_2.

The pure gas is then methanated on a nickel catalyst at $400^{o}C$. After passing through a waste heat cooler, in which the water produced by methanation condenses out, it is injected into the SNG mains network.

VIII. Construction and development

The process for the steam gasification of hard coals using nuclear process heat is being developed by Bergbau-Forschung GmbH, Essen.

References

(1) Kohlevergasung mit nuklearer Prozesswärme (Coal gasification using nuclear process heat), Düsseldorf, 1974:
A study carried out for the Nordrhein-Westfalen Regional Ministry for Economic Affairs, Small Businesses and Trade by a consortium formed by Gesellschaft für Hochtemperaturreaktor-Technik mbH Bensberg; Bergbau-Forschung GmbH, Essen; Rheinische Braunkohlenwerke AG, Cologne; and Steag AG, Essen.

(2) K.H. van Heek, H. Jüntgen and W. Peters, Erdöl und Kohle 26 (1973), 701

(3) H. Jüntgen, K.H. van Heek and J. Klein, Chem. Ing. Techn. 46 (1974), 937

(4) H. Krüger, K.H. van Heek and H. Jüntgen, Gas- und Wasserfach - Gas/Erdgas 115 (1974), 538

(5) H. Jüntgen, VDI - Ber. 224 (1974), 13

(6) Prototypanlage nukleare Prozesswärme (PNP) (Prototype plant for using nuclear process heat), December 1976:
Progress report at the end of the design phase, prepared by the consortium formed by Bergbau-Forschung GmbH, Essen; Gesellschaft für Hochtemperaturreaktor-Technik mbH, Bensberg; Hochtemperatur-Reaktorbau GmbH, Mannheim; Kernforschungsanlage Jülich GmbH, Jülich; and Rheinische Braunkohlenwerke AG, Cologne.

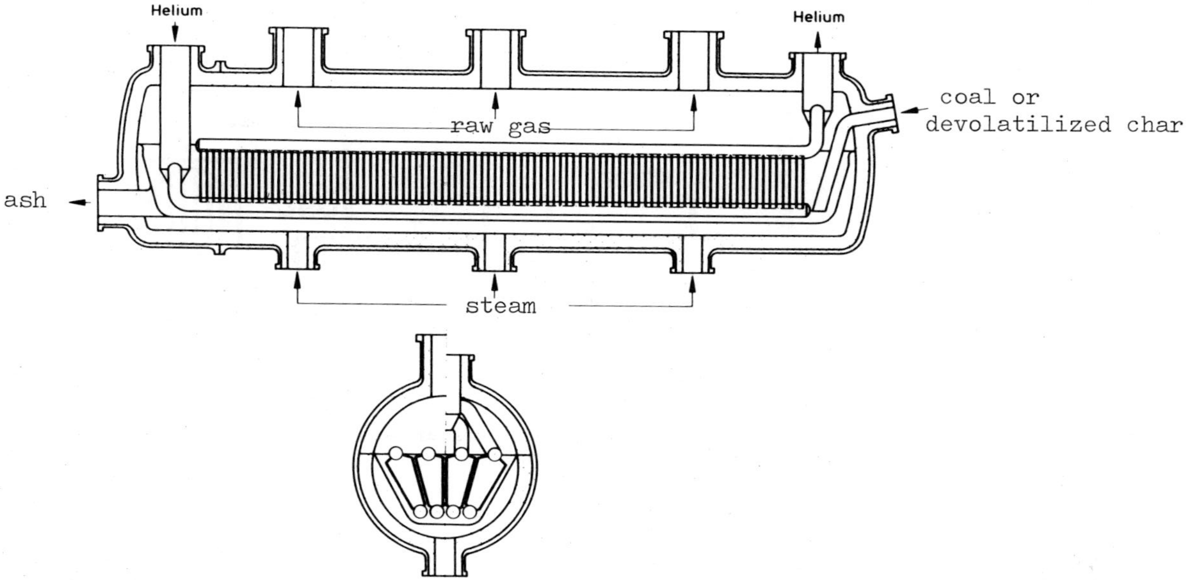

Figure 38 Sketch of a gasification reactor

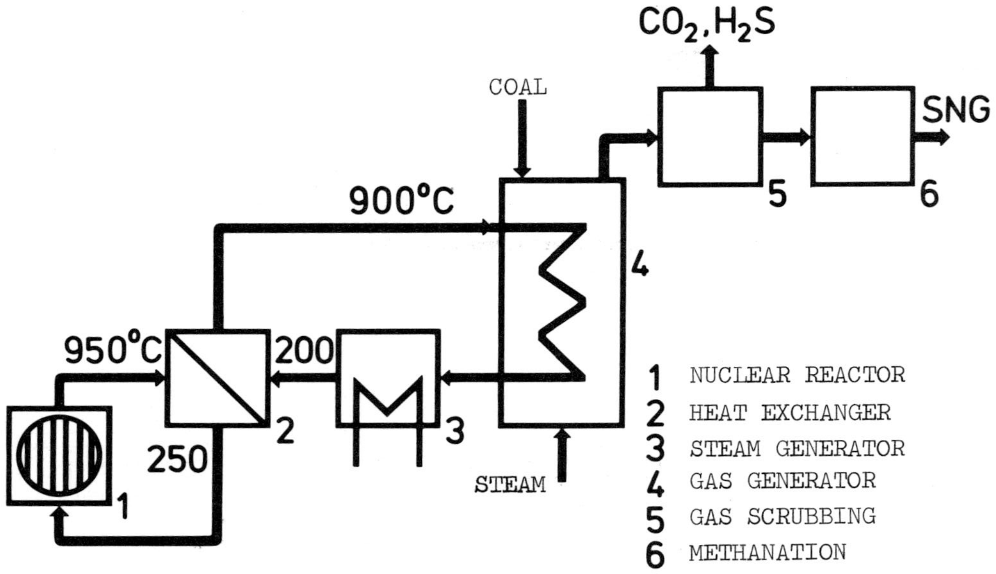

Figure 39 Flow-sheet for the steam gasification of hard coal

3. Hydrogasification of lignite

In this section descriptions are given of two variants of the hydrogasification of Rhenish lignite using nuclear process heat. In the first of these lignite is converted to methane in exothermic reaction with hydrogen. Part of the methane is then reacted with steam in endothermic reaction, using nuclear process heat, to give carbon dioxide and hydrogen. The hydrogen is used as gasifying medium.

The second variant is based on the hydrogasification of lignite with a mixture of hydrogen, carbon monoxide and steam. The mixed product gas contains methane, which is separated out. The remainder is recirculated after the addition of more steam. The production of hydrogen by the heterogeneous and homogeneous water-gas reaction is thus combined with hydrogasification in a single reactor. The heat required for the reaction is provided by heating the gas used for gasification in a heat exchanger with nuclear process heat (hydrogasification heater).

The two processes are described in turn below.

Indirect coupling of nuclear heat via steam reforming of methane in a tube furnace

I. Gasification system

Hydrogasification of dried lignite at 80 bar in a fluidized-bed reactor.

II. Performance

A study carried out for the Nordrhein-Westfalen Regional Ministry of Economic Affairs, Small Businesses and Trade gave a conceptual design for a large plant consisting of 12 gasifiers each of 5 m diameter; 2228 t of raw lignite would be gasified per hour, producing about 2 million m^3(iN)/h of dry raw gas.

The following figures refer to a single gasifier:

Throughput:	185.6 t raw lignite/h
Raw gas production:	165.6 x 10^3 m^3(iN)/h
Dimensions: diameter	5 m
height (including drying stage and char cooler)	ca 16.5 m

The study is based on the results of laboratory-scale experiments. Meanwhile, a semi-technical experimental plant with a throughput of 100 kg C/h is being operated by Rheinische Braunkohlenwerke AG. No data are available for this plant yet.

III. Feed coal requirements

a. Rank

The process has been designed for Rhenish lignites.

Elementary analysis:	C	26.3	wt %
	H	2.0	"
	O	9.7	"
	N	0.4	"
	S	0.2	"
	Ash	2.4	"
	H_2O	59.0	"
Calorific value:		10.5 MJ/kg (2500 kcal/kg)	

Non-caking hard coals will also be used later.

b. Particle size

Tests showed that the particle size for fluidized-bed gasification of lignite should be less than about 1 mm. To allow for shrinkage on drying, the coal is therefore pulverized down to a particle size < 1.5 mm.

c. Ash

No information is given on the permissible limits for ash content, since the study related specifically to the use of Rhenish lignites.

d. Coal pretreatment

The raw lignite is first ground to a particle size of 0-10 mm and is then dried in two stages. In the first stage it is predried to a water content of 41%

in tubular driers. Low-pressure steam is used here, which condenses in the driers. This steam is obtained from the waste heat boilers of the gasification plant.

After this preliminary drying the coal is repulverized down to its final particle size of 0-1.5 mm. This is followed by the second stage of drying in a pressurized fluidized bed fed with the raw gas, which leaves the gas producer at 850°C. In this way the coal is dried to a residual moisture content of 4% and preheated to 300°C; at this temperature it is transferred to the gasifier.

IV. Raw gas composition (dry, tar-free basis)

H_2	62.9 vol %
CO	4.3 vol %
CO_2	0.73 vol %
CH_4	31.5 vol %
C_2H_6	0.08 vol %
N_2	0.32 vol %
H_2S	0.13 vol %

Calorific value: 21.2 MJ/m^3(iN)
(= 5056 $kcal/m^3$(iN))

V. Description

After the initial grinding the raw lignite is first dried to a residual water content of 41%; after fine pulverization it is then conveyed pneumatically to the storage bunkers. From there it passes into pressurized lockhoppers, sealed by inert gas (CO_2), and then into the fluidized bed, where direct contact with the raw gas from gasification brings it to its final moisture content of 4% and to a temperature of 300°C, at which it enters the gasifier.

It falls through an overflow duct into the gasifier, which is likewise constructed as a fluidized bed. The gasifying medium consists of 98% hydrogen and the gasification temperature is 850°C. The residence time in the fluidized bed is envisaged as 30 minutes, which should suffice to gasify 63.5% of the carbon.

The residual char passes through an overflow duct into the char cooler, likewise constructed as a fluidized bed, in which the gasifying medium, already preheated to 300°C, is raised to 380°C by direct contact with the char. Finally, any non-gasified char is discharged through lockhoppers sealed, like the coal lockhoppers, with inert gas. It can be put to further use in, for example, making synthesis gas by steam gasification, under-boiler firing, or metallurgical processes.

The hydrogenation gas is first preheated to 300°C in waste heat boilers and then raised to 380°C in the char coolers; at this temperature it enters the gasification reactor. The exothermic reaction raises the temperature of the fluidized bed to 850°C. The raw gas leaves the gasifier at this temperature and is cooled to 300°C in the second coal-drying stage. It then passes through a cyclone for removal of entrained coarse dust. The remaining waste heat, together with waste heat from the carbon monoxide conversion, serves to preheat the hydrogenation gas, to produce low-pressure steam for the first coal-drying stage, to preheat the boiler feed water, and to operate the gas scrubber (reboiler).

The fine dust is removed from the raw gas by a water scrubber, which follows the first waste heat boiler. This reduces the temperature of the gas from 250°C to 218°C.

Since the subsequent stages of the process (low-temperature fractionation and catalytic reforming of methane) require gas of high purity, the residual content of CO_2 is reduced to about 10 ppm and that of H_2S to about 1 ppm in an Amisol unit. The raw gas is first washed with methanol at 10°C in order to remove any naphtha components. The subsequent Amisol wash is carried out at 40°C, and the acid gases obtained are treated in a Claus kiln for sulphur recovery.

After this, the purified gas is cooled to -52°C, combined with the gas from the conversion unit, and subjected to low-temperature fractionation. First, the mixed gas is passed through molecular sieve adsorbents in order to remove residual CO_2 and traces of methanol vapour. Then it is cooled to -180°C, which condenses out all components except hydrogen. The hydrogen is 98% pure and serves as the gasifying medium. The liquefied mixture is fractionally evaporated: one fraction, rich in hydrogen and carbon monoxide, is returned to the conversion unit; the other fraction, rich in methane, passes on to further purification in a fractionating column. The nitrogen-rich mixture removed there is emitted as off-gas.

Approximately 55% of the methane (380×10^3 m^3(iN)/h) is injected into the public mains network as SNG. The remainder, mixed with steam, is catalytically reformed to hydrogen and carbon monoxide at 40 bar and about 800°C in a tube furnace heated by nuclear process heat. The resulting gas is converted in two stages at about 400°C, the H_2/CO fraction from the low-temperature fractionation being added before the second stage. After passing through the waste heat exchanger the gas is compressed to 77 bar and freed from CO_2 by Rectisol washing: the CO_2 is used as the inert sealing gas in the lockhoppers, and the remaining gas is combined with the purified gasification gas and fed into the low-temperature fractionation plant.

For starting up the gasifier, to begin with oxygen is added to the hydrogen before it enters the fluidized bed, so that part of the hydrogenation gas is burned and the fluidized bed is heated by the combustion gases. Once the hydrogasification reaction has set in, this addition of oxygen can be cut down and finally stopped altogether.

VI. Production of gasifying medium

Hydrogen: The hydrogen used for gasification (98% pure) is obtained by the catalytic reforming with steam of part of the methane produced. This reaction, which is fairly strongly endothermic, is carried out in tubular reformers which are directly heated by hot helium in the vessel of the high-temperature nuclear reactor.

VII. Technical data (for gas production)

a. Carbon conversion efficiency: 63.5%

b. Gasification efficiency: Not applicable (incomplete gasification)

c. Thermal efficiency (including treatment of gas): 88.7% in terms of calorific values

d. % steam decomposition: Not applicable (hydrogasification)

e. Inputs:

Steam (41 bar/322°C):	440	kg/t raw lignite
	0.493	kg/m^3 dry raw gas
	23.3	kg/GJ in raw gas
Hydrogen: (98%, remainder CH_4)	550	m^3(iN)/t raw coal
(from the integral methane reformers)	0.617	m^3(iN)/m^3(iN)dry raw gas
	29.1	m^3(iN)/GJ in raw gas
Coal:	1.12	kg raw lignite/m^3(iN) dry raw gas
	53.0	kg raw lignite/GJ in raw gas

f. Steam/oxygen ratio: Not applicable (hydrogasification)

g. Products:

Gas yield:	892 m^3(iN) dry raw gas/t raw coal 170 m^3 SNG (net)/t raw coal
Residual char:	122 kg/t raw coal

Analysis of residual char:

C	78.4	wt %
H	1.13	"
O	0.35	"
N	0.31	"
S	0.26	"
Ash	19.6	"

Calorific value: 28.2 MJ/kg (= 6730 kcal/kg)

h. Gasification pressure: 80 bar

Gasification temperature: 850°C

Exit gas temperature: 850°C

State of gasifying medium: Hydrogen, 98% pure, preheated to 380°C

i. Water consumption: The water consumption of the gas production plant is estimated at 520 m^3/h. This is calculated as the difference between the amount of steam delivered from the steam generator and the amount of condensates returned.

The water used for scrubbing comes solely from the condensates.

	800 m^3/h of aqueous effluent are expected. The extent to which treated effluent can be recycled remains an open question.

j. Other energy consumption:

Process heat from high-temperature nuclear reactor (900°C)	54.9	MJ/GJ in raw gas
	1.042	GJ/t raw coal
Electricity	4.97	kWh/GJ in raw gas
	94.3	kWh/t raw coal
Low-pressure steam (13 bar/233°C)	10.0	kg/GJ in raw gas
	189	kg/t raw coal
For driving compressor turbines:		
(low-pressure steam, 39 bar/360°C)	6.16	kWh/GJ in raw gas
	117	kWh/t raw coal

k. Gas purification

As already described in Section V, the gas first undergoes coarse dust removal in a cyclone unit and is then washed with water to remove fine dust. The next step is an Amisol wash (preceded by washing with methanol), which largely removes the acid gases. The methanol wash serves to remove any higher hydrocarbons which may be present. The acid gases are treated in a Claus kiln for sulphur recovery.

The pure gas mixture is then fractionated at a low temperature. The hydrogen is recirculated as gasifying medium, while the CO-rich fraction is converted and fed back into the low-temperature fractionation plant. The methane-rich fraction is re-fractionated, and a nitrogen-rich mixture is separated as off-gas.

Approximately half the methane is taken off as product gas; the remainder is catalytically reformed with steam to hydrogen and carbon monoxide in a tubular furnace at 800°C. The carbon monoxide is then converted immediately, while the hydrogen is used for hydrogasification of the coal.

VIII. Construction and development

The process for hydrogasification of lignite is being developed by Rheinische Braunkohlenwerke AG, Cologne.

References

(1) Kohlevergasung mit nuklearer Prozesswärme (Coal gasification using nuclear process heat), Düsseldorf, 1974:
Study carried out for the Nordrhein-Westfalen Regional Minister of Economic Affairs, Small Businesses and Trade by a consortium formed by Gesellschaft für Hochtemperaturreaktor-Technik mbH, Bensberg; Bergbau-Forschung GmbH, Essen; Rheinische Braunkohlenwerke AG, Cologne; and Steag Aktiengesellschaft, Essen.

(2) R. Hüttner and H. Teggers, Braunkohle-Wärme-Energie 23 (1971), 134

(3) P. Speich, VGB Kraftwerkstechn. 54 (1974), 438

(4) H. Teggers, Gas- und Wasserfach - Gas/Erdgas 115 (1974), 532

(5) Prototypanlage nukleare Prozesswärme (PNP) (Prototype plant for using nuclear process heat), December 1976:
Progress report at the end of the design phase, prepared by the consortium formed by Bergbau-Forschung GmbH, Essen; Gesellschaft für Hochtemperaturreaktor-Technik mbH, Bensberg; Hochtemperatur-Reaktorbau GmbH, Mannheim; Kernforschungsanlage Jülich GmbH, Jülich; and Rheinische Braunkohlenwerke AG, Cologne.

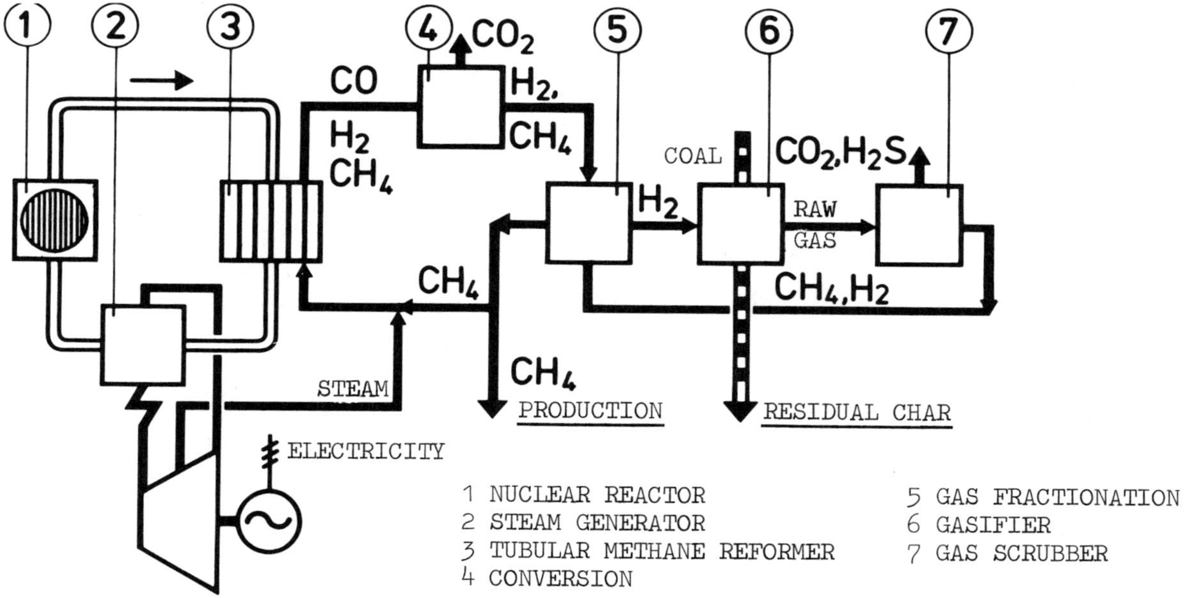

Figure 40 Flow-sheet of a process for hydrogasification of coal using process heat from a high-temperature nuclear reactor

Coupling of nuclear heat via pre-heating of the gasifying medium

I. Gasification system

Hydrogasification of dried lignite in a fluidized-bed reactor at 80 bar with an externally heated gasifying medium consisting of hydrogen, carbon monoxide and steam.

II. Performance

The study gives a conceptual design for a plant consisting of 12 gasifiers each 5 m in diameter which would provide a throughput of 2515 t of Rhenish lignite/h and a dry raw gas yield of 2.04×10^6 m^3 (iN)/h; 331 t of residual char/h are also produced.

The following figures refer to a single gasifier:

Throughput:	209.6 t raw lignite/h
Raw gas production:	169 700 m^3(iN)/h of dry raw gas

The design is based on test data from a small continuous experimental plant at the Institute of Gas Technology, Chicago, USA.

III. Feed coal requirements

The type of coal and its pretreatment are the same as for the hydrogasification process described in the preceding section.

IV. Raw gas composition (dry, tar-free basis)

H_2	37.7	vol %
CO	15.1	vol %
CO_2	15.7	vol %
CH_4	20.9	vol %
N_2	10.5	vol %
H_2S	0.14	vol %

Calorific value: 15.1 MJ/m^3(iN)
(= 3590 $kcal/m^3$(iN))

V. Description

The pretreatment of the coal is the same as that described for the preceding process.

The dried coal passes through an overflow duct into the actual gasification reactor, likewise a fluidized bed, where it meets a hydrogenation gas having the following composition:

H_2	33.0	vol %
CO	13.0	vol %
CH_4	0.7	vol %
N_2	9.0	vol %
H_2O	44.3	vol %

The gas enters at 900°C and 60% of the lignite would be gasified; the operating pressure is 80 bar.

After passing through the fluidized-bed drier the raw gas is cooled from 300°C to 250°C in the waste heat boiler and is freed from dust in a cyclone followed by a water scrubber. The heat still remaining in the gas is used for the Rectisol unit, for generating low-pressure steam, for the preliminary drying of the coal and for preheating boiler feed water.

Because of the relatively high CO_2-- content of the gas, further purification is carried out by three stages of Rectisol washing. The first stage serves to remove mainly naphtha and other higher hydrocarbons, together with residual water vapour, while the next two free the gas from CO_2 and sulphur compounds.

The acid gases obtained contain comparatively small amounts of sulphur compounds, and their further treatment for recovery of elementary sulphur therefore involves an oxidizing wash (Giammarco process). The carbon dioxide will be used as a sealing gas for the coal and char lockhoppers.

The pure gas is separated by low-temperature fractionation into hydrogen, carbon monoxide and methane fractions. The methane fraction is injected at 70 bar into the natural-gas mains network, while the other two fractions are combined and fed back into the hydrogasification cycle together with the steam required. No conversion unit is needed in this process.

Thus, the essential difference between this process and the hydrogasification process described in the preceding section is that the hydrogen required for methane formation is produced in the gas generator itself by gasification with steam. This avoids any need for a conversion unit and methane reformer, but the hydrogasifying medium has to be preheated to a far higher temperature.

VI. Production of gasifying medium

Hydrogen:	Hydrogen is produced in the gasifier by the steam gasification of coal.
Steam:	High-pressure steam at 82 bar is produced with the hot helium from a high-temperature nuclear reactor. The temperature range from 950°C to 800°C serves to preheat the hydrogenation gas, while the range from 800°C to 250°C is used for steam generation.

VII. Technical data (gas production only)

a. Carbon conversion efficiency:	60%	
b. Gasification efficiency:	Not applicable (incomplete gasification)	
c. Thermal efficiency:	93.5% in terms of calorific values	
d. % steam decomposition:	25.3%	
e. Inputs:		
Steam: (80 bar, 535°C)	326	kg/t raw coal
	0.403	kg/m^3(iN) dry raw gas
	112	kg/Gcal in raw gas
Coal:	1.24	kg raw lignite/m^3(iN) dry raw gas
	344	kg raw lignite/Gcal in raw gas
f. Steam/oxygen ratio:	Not applicable	
g. Products:		
Gas yield:	809.5	m^3(iN) dry raw gas/t raw lignite
	165	m^3 SNG(net)/t raw lignite

Residual char: 131.6 kg residual char/t raw lignite

Analysis of residual char:

C	79.74 wt %
H	1.15 "
O	0.36 "
N	0.31 "
S	0.26 "
Ash	18.18 "

Calorific value: 28.7 MJ/kg (= 6850 kcal/kg)

h. Gasification pressure: 80 bar

Gasification temperature: 800°C

Exit gas temperature: 800°C

State of gasifying medium: hydrogenation mixture, 82 bar, 900°C

i. Water consumption: 922 m^3/h (with maximum recycling of condensed water)

j. Other energy consumption:

Process heat from high-temperature nuclear reactor (900°C):	0.318 GJ/Gcal in raw gas 0.923 GJ/t raw lignite
Electricity:	39.4 kWh/Gcal in raw gas 114.5 kWh/t raw lignite
Low-pressure steam: (8 bar/234°C)	24.6 kg/Gcal in raw gas 71.6 kg/t raw lignite

k. Gas purification:

The gas is first freed from dust, etc, by cyclones followed by water scrubbers, as in the last process described. There then follows a three-stage Rectisol wash for removing hydrocarbons and acid gases. Once purified in this way, the gas satisfies the requirements for low-temperature fractionation, in which the methane formed in the gasifier is separated out. The remaining fractions, rich in hydrogen and CO, are recombined and recycled as hydrogasifying medium: no conversion is necessary. Because of their low sulphur compound content, the acid gases are treated in an oxidizing wash (Giammarco process) for recovery of elementary sulphur.

VIII. Process development

This process has been proposed as a variant of the last process described, again as part of the study carried out for the Nordrhein-Westfalen Regional Ministry of Economic Affairs, Small Businesses and Trade.

However, no known material offers the properties required for the materials to be used in the construction of the preheater for the hydrogasifying medium; similarly, no equipment yet exists for handling hot gas under the required limiting conditions. For these reasons the process is not at present being developed in this form, despite its great technological advantages.

References

(1) Kohlevergasung mit nuklearer Prozesswärme (Coal gasification using nuclear process heat), Düsseldorf, 1974:
Study carried out for the Regional Minister for Economic Affairs, Small Businesses and Trade, Nordrhein-Westfalen, by a consortium formed by Gesellschaft für Hochtemperaturreaktor-Technik mbH, Bensberg; Bergbau-Forschung GmbH, Essen; Rheinische Braunkohlenwerke AG, Cologne; and Steag Aktiengesellschaft, Essen.

(2) R. Hüttner and H. Teggers,
Braunkohle-Wärme-Energie 23 (1971), 134

(3) P. Speich, VGB Kraftwerkstechn. 54 (1974), 438

(4) H. Teggers, Gas- und Wasserfach - Gas/Erdgas 115
(1974), 532

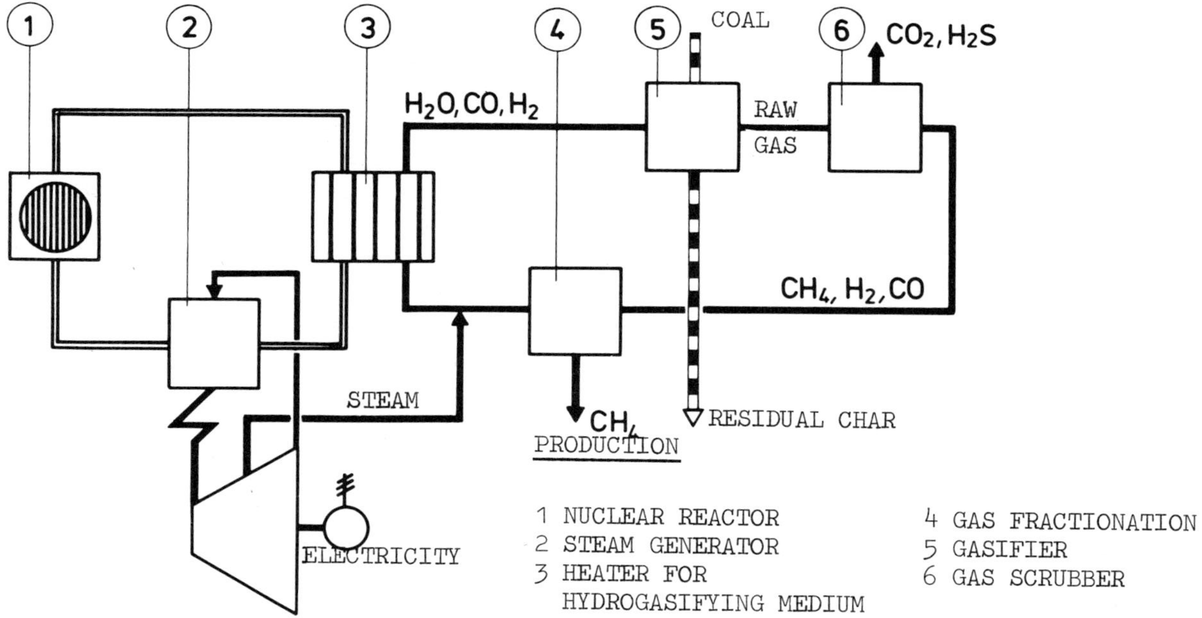

Figure 41 Flow-sheet of a process for hydrogasification of coal using process heat from a high-temperature nuclear reactor

4. Hydrogasification of hard coal

Similar experiments are planned, but have not yet been published.

E. Underground gasification

The object of underground gasification is to convert the coal directly into gas in situ, without mining it. To do this, suitable sections of the coal seam to be gasified have to be prepared in advance. This preparation can be carried out either from above ground by means of specially sunk boreholes, or from underground by means of shafts and mine workings. The former of these procedures is used today.

This method of extracting coal has several advantages. It obviates both the arduous working conditions to which miners are subjected in underground mining and, at the same time, the surface damage caused by opencast mining. In addition, the coal is converted in situ into gas, which is more simply handled than solid fuel. Furthermore, the conversion involves practically no damage to the biosphere. Coal with a high ash content can be used, and also coal rich in sulphur, since the removal of sulphur from the product gas is a well-established technology. Not least, the technique of underground gasification lends itself to extensive automation.

The first idea of using this technique for exploiting coal deposits was developed in England in the middle of the last century. However, it was not until the 1930s that the first practical applications took place, in the USSR. Since then installations for underground gasification have been in operation in the USSR up to the present day, though with some interruptions caused by damage during the war. After 1945 interest in underground gasification was aroused in a number of other countries. Field experimentation took place in the UK, France, Belgium, Poland, Czechoslovakia and the USA. These tests were quite soon abandoned, mainly because of unconvincing results. The availability of cheap oil and large finds of natural gas also reduced the incentive for underground gasification. By 1960 all attempts at development had been abandoned. The Russian plants installed up to that date continued to operate, but the rate of development clearly slackened. According to present information the construction of additional installations which were planned for the early 1960s did not take place.

The installations set up in the USSR are at present the largest operations of this kind anywhere in the world. They are based on the gasification of both hard coal (Donets and Kuznetsk regions) and lignite (deposits near Moscow, and Tashkent in Soviet Central Asia).

Since the beginning of the 1970s there has been a revival of interest in underground gasification, first in the USA and then in other countries such as Belgium and the Federal Republic of Germany. In the USA the change in attitude has been prompted by the impending bottleneck in the supplies of natural gas. It is hoped that underground gasification will contribute to US self-sufficiency in energy. In addition, the successful introduction of underground gasification might make it possible under some circumstances to exploit coal deposits which could not hitherto be mined economically by conventional means, thus increasing reserves for the future: for example, deep-lying coal deposits in the western USA might be exploited profitably.

This last argument for more efficient utilization of existing resources has also had effect in Europe. Underground gasification was seized upon in Belgium as virtually the only possible way of exploiting the extensive Belgian deposits which cannot be mined economically by conventional means because the seams are not thick enough.

This method of extraction is also of fundamental interest in the Federal Republic of Germany in the light of the conditions governing its resources of hard coal. It is estimated that the total reserves of hard coal amount to about 230 000 million t, of which about 24 000 million t are suitable for mining.

This means that, on present figures, about 90% of the total resources are inaccessible to conventional mining methods because they are too deep or the seams are too thin. The development of an unconventional method of extraction which makes it possible to exploit the coal without underground mining is therefore of great potential interest.

Almost all existing operations and most projected ones are designed to produce a lean gas with a calorific value of between 3.5 and 5 MJ/m^3(iN) (800-1200 $kcal/m^3$(iN)), which cannot be transported economically over even moderate distances. German plans therefore include the direct production of SNG by gasification with hydrogen (Variable-Pressure process - RWTH Aachen (Nordrhein-Westfalen Technical University)).

However, a number of problems remain to be solved before underground gasification can be used commercially on a large scale. This technique suffers from the disadvantage that its operating results are heavily dependent on geological and hydrogeological conditions; moreover, the nature of this dependence cannot be expressed in a quantitative form. This makes it difficult to transpose operating data from one deposit to another for planning purposes. Furthermore, in contrast to gasification in a generator above ground, in this case the reaction space is probably not gas-tight and its dimensions vary as gasification proceeds. As a result, the conditions for steady operation are also changing continuously. This means that, particularly in gasification with air, oxygen and steam, or mixtures of these gases, both the yield and quality of the product gas will often vary under some circumstances. Tests to date have sometimes shown that both the quality of the gas and the gasification efficiency decrease as operations proceed. However, a considerable degree of constancy in output and quality would seem necessary for successful industrial operation on a large scale. New concepts therefore need to be developed to eliminate these disadvantages, either directly in the process itself, or indirectly by, for instance, interconnecting several gasification fields.

Almost all the processes proposed today suffer from a common difficulty. In contrast to gasification above ground, with gasification in situ it is not really possible to control the movement of the gasifying medium in relation to the fuel. The fuel (the coal seam) is fixed, while the stream of gasifying medium is almost uncontrollable once it has been admitted to permeable channels: it will be only partly directed towards the coal face, and may reach the gas outlet without passing through the reaction zone. The degree of conversion is thus determined more by chance than by any technological relationships. In consequence, the quality of the gas may be impaired by reaction of the gas first produced with free oxygen, and it is very difficult to control the movement of the reaction front.

A partial step towards the effective control of underground gasification has been proposed recently. This consists of burning out highly permeable channels beforehand ("back-burning") through which the reaction front and the stream of gasifying medium pass in opposite directions, for example between two boreholes. However, experience to date suggests that this process cannot be used to produce and maintain extensive reaction zones, since the structure of coal seams is inhomogeneous as regards permeability to gases and favours break-through at scattered points, as shown by an experiment in the USA.

A way of avoiding this difficulty is embodied in the concept of the Thick Packed Bed process (Lawrence Livermore Laboratory, Livermore, California), in which a thick coal seam, previously fractured by explosive charges, acts as a kind of packed-bed reactor and is gasified in situ from its roof to its floor.

Yet another problem hinders the extension of underground gasification to operation at considerable depths, especially in Europe. To date, operations have been carried out predominantly at depths of about 100 m, and only in a few test experiments has a depth of 300-400 m been reached. However, at great depths, eg 1000 m or more, the design of the process and the necessary number of boreholes (the method normally used today) become determining factors from an economic point of view. Under these circumstances, minimization of the number of boreholes appears to be the vital criterion in choosing a suitable process.

Although the introduction of underground gasification on a large industrial scale thus involves a number of major difficulties, especially for deep-lying deposits, the advantages which this technique may offer are more than enough to justify studying the feasibility of its application on a large industrial scale to the deep-lying deposits in the Federal Republic of Germany. This task is included in the general energy research programme of the Bundesministerium für Forschung und Technologie (Federal Ministry for Research and Technology).

As an aid to better understanding of underground gasification and the current development status of this technology, the following sections contain a comparison of techniques and results for some selected installations. It should be borne in mind that, for the reasons already stated, the results are not directly transferable and apply in the first instance only under the specific conditions of the plant to which they refer; they can, however, be regarded as reference values.

Figures on manpower requirements and the resultant labour productivity in the Soviet installations which in some cases are quoted in the literature (the available data refer to 1963) are not to be taken as representative of modern installations such as those likely to be constructed in the USA or Europe. The Russian installations were constructed largely for research purposes, and therefore had well above average manning levels. As a result, labour productivity in the smaller plants, for example, was at times only about 330 000 m^3(iN) of product gas per man-year or, in terms of standard fuel (assuming a calorific value of H_o = 1100 kcal/m^3(iN)), about 30-50 tce per man-year. In larger plants, designed more for industrial operation, values of up to nearly 200 tce per man-year were reached. It should be noted that these figures refer to energy in the form of gas, and also to the total number of employees in the plant. By comparison, productivity in conventional mining in the Federal Republic of Germany in 1975 was on average 700 tce per underground worker.

A description is given of the basic operating principles and results of the Angren installation (Uzbekistan, USSR), since this is the largest so far operated anywhere in the world. In this plant a thick, level seam of lignite is gasified in situ. The largest plant for the gasification of hard coal is installed on a deposit at Yuzhno-Abinsk (Kemerovo, Kuznetsk Basin, USSR) consisting of moderately or steeply inclined seams. As a third example, results are given for a large-scale experiment carried out in 1976 at Hanna (Wyoming, USA). Only lean gas has been produced in all three cases. It is not yet possible to give the results of underground tests designed to produce gas of medium calorific value (Thick Packed Bed process, USA) or SNG (Variable Pressure process, RWTH Aachen), since only preliminary laboratory investigations have been carried out for the latter, while no authoritative figures are available for the first field tests of the Thick Packed Bed process.

Operational data for the Angren and Yuzhno-Abinsk installations (both USSR) and results of experiments at Hanna (USA)

I. Gasification system (layout of working section)

A. Angren (USSR)

A thick, level seam of lignite is gasified with air. Rows of boreholes are drilled down into the coal seam at an angle smaller than the angle of slope of the overburden. Initially, these are used for injecting the gasifying medium; then, when the reaction zone in the seam has moved above the foot of the borehole, they are also used for drawing off the gas. In order to ensure a sufficient flow of gas between the ends of the boreholes, narrow channels readily permeable to gas are burned in the coal by injecting compressed air before gasification proper begins. The boreholes are spaced at a distance of up to about 50 m (see Figure 42).

B. Yuzhno-Abinsk (USSR)

Steeply dipping seams (0-67°) are gasified by injecting air. Boreholes for extracting the product gas are sunk into the inclination of the seam to the required depth (about 300 m). The boreholes for injecting the gasifying medium are made through the bedrock. When they reach the seam they are bent up into it. The distance between the boreholes for extracting gas and for injecting air is between 20 and 35 m (see Figure 43).

C. Hanna (USA)

All the tests at Hanna have been carried out on a level, thick seam of lignite. In an investigation between May and July 1976 four vertical boreholes were sunk at the corners of a square with 18 m sides.

Between each pair of boreholes readily permeable channels were burned through the bottom third of the seam. Actual gasification took place after the channels had been completed. The gasifying medium was injected through one borehole; it then passed along the channel, reacting with the coal, and was taken off through the second borehole (see Figure 44).

II. Performance

For the Russian installations mean values are given for the year in which (as far as is known) the best results were obtained. The figures given for the Hanna tests are mean values over the period May-July 1976.

	A (1966)	B (1960)	C (1976)	
Gas production or	161	41	5-13	$10^3 m^3$(iN)/h
Coal gasified	ca 55	ca 9	ca 4	t/h
Corresponding to an electricity generation calculated as:	ca 60	ca 20	ca 3-8	MW

III. Characteristics of coals and structure of deposits

1. Characteristics of coals

At Angren a deposit of lignite is used for gasification, and in Yuzhno-Abinsk a deposit of hard coal. Seam No 1 used in the Hanna experiments also consists of a lignite (vitrain). The following analyses give average values for the composition of the coals.

Coal analyses (untreated)

	A	B	C
Direct analyses			
Moisture, wt %	31	6.0	9.5
Volatiles, wt %	ca 23	ca 40	32.6
Fixed carbon, wt %	ca 35	ca 50	34.1
Ash, wt %	11	4.9	23.8
Elementary analyses (wt %)			
C	43.3	75.0	50.0
H	2.3	4.9	4.1
O	10.9	7.1	11.2
N	0.5	2.1	1.3
S	1.5	0.3	0.7
Calorific value H_o (kcal/kg)	4000	7500	5080
(MJ/kg)	16.7	31.5	21.3

2. Structure of deposits

		A	B	C
Test depth	m	100-200	100-300	84
Inclination	degrees	5-15	60-67	6-9
Mean thickness of seam	m	9	2.5	9
Range of thicknesses	m	(1.5-21)	(1-9)	(8-9)
Country rock		clay and sandstone 10-30 m thick	clay and soft sand-stone 3-5 m thick	sandstone shale

IV. Raw gas

Composition (dry basis):

		A	B	C
H_2	vol %	17-20	17-18	10-19
O_2	vol %	0.3-0.4	0.2	--
CO	vol %	4-8	17-18	11-20
CO_2	vol %	17-20	9-10	8-15
$CH_4 + C_nH_m$	vol %	2.0	2.0	2-6
N_2	vol %	50-58	56-57	46-58
H_2S	vol %	0.4-0.5	0.2	ca 1
Calorific	kcal/m^3(iN)	ca 1000	ca 1150	ca 1500
value H_o	MJ/m^3(iN)	ca 4.2	ca 4.8	ca 6.3

The values for other constituents of the raw gas are known only for phase I of the Hanna experiment. Moisture contents of between 8 and 28 wt % were measured in the raw product gas. Its tar content lay between 1 and 1.5 wt %. The amount of dust in the gas fluctuated between 5 and 870 mg/m^3(iN).

The fact that, overall, the results given under C (Hanna) are better than average is probably at least partly due to the relatively short period of operation and the small scale of the experiment.

The product gas from the Russian installations is used for bottom firing, for space heating, and also in industrial plants and in the power industry. The product gas from the Hanna tests was flared off.

V. Description

Figures 42, 43 and 44 (at the end of the chapter) show the basic layout of the gasification fields in the installations described above. They also illustrate the course of the process. It should be noted that differences in the geological conditions surrounding deposits mean that operating conditions in underground gasification will usually vary widely. The gasifying medium flows through a system of pipes into the bore-holes, from which it is forced into the seam. After ignition a reaction zone is formed in the seam, in which most of the reaction between the gasifying medium and the coal takes place. This zone may consist of a slowly moving, more or less linear front (Angren), or of a highly porous channel permeable to gas, in which gasification proceeds along the walls. In steeply inclined seams this channel will move upwards along the seam without changing its geometric form, so that in a single gasification section fairly constant results may be obtained over a reasonably lengthy period.

On the other hand, when gasification in channels is applied to level seams the channels will clearly become wider as operations proceed and the coal is burned away. This leads to deteriorating operational results as a result of the diminishing yield and the increased possibility of aftercombustion of the product gas with oxygen that has not been consumed in gasification. It is obviously advisable to adjust the supply of gasifying medium continuously to the changing dimensions of the cavity which has been burned out: this will depend, among other things, on the duration of gasification operations, the thickness of the seam, the quantity and properties of the ash, and the stability of the terrain. The amount of gasifying medium required will therefore increase continuously with the time for which a given gasification field has been worked.

As demonstrated in the Hanna tests, the expansion of

the cavity, which is disadvantageous for gasification purposes, can be partially compensated if the gas pressure in the deposit is raised by throttling down the gas take-off, which also helps to prevent ingress of water.

The product gas is extracted from the deposit through the take-off boreholes, and after appropriate purification by conventional procedures, it is ready for use.

In addition to the gasification field itself, which corresponds conceptually to the generator in a surface gasification plant, auxiliary facilities are required for preparing the gasifying medium (for example compressors, or an oxygen plant) and for treating the gas according to its intended use. Also, continuous industrial operation necessitates the constant availability of machines and manpower for further drilling.

VI. Production of gasifying medium

In all three installations gasification is carried out with compressed air produced in a compressor plant.

VII. Technical data

Efficiency

The coal utilization efficiency is defined as the proportion of the coal in the seam which is converted by reaction with the gasifying medium. The extent to which the available coal is utilized is very closely governed by the layout of the gasification field, and hence on the array of boreholes. The gasification efficiency does not refer to the total amount of coal present, but only to that proportion of it which is utilized.

	A	B	C
a. Coal utilization efficiency, %	65-70	55-100	not stated
b. Gasification efficiency, %	55-65	55-60	ca 50-80
c. Thermal efficiency	not stated	not stated	not stated
<u>Inputs</u>			
Gasifying medium 10^3 m^3(iN)/h	not stated	not stated	2.8-7.2
<u>Products</u>			
Gas yield (referred to gasifying medium) m^3(iN)/m^3(iN)	1.1-1.3	not stated	1.7
<u>Gasification pressure</u>			
Injection pressure, bar	5	5	4.5-9
Pressure of product gas, bar	not stated	not stated	2-5

An upper pressure limit is fixed by factors that include the lithostatic and hydrostatic pressure of the deposit. In gasification near the surface, only low average gasification pressures can be used, because of the relatively high permeability of the overburden to gas: otherwise, gas losses (which in any case can not be avoided altogether) would reach a prohibitive level. In gasification at great depth, on the other hand, the superimposed pressure makes the country rock sufficiently dense to allow a higher range of pressures to be used. The proximity of an aquifier may in fact make it necessary to maintain a certain pressure level in order to prevent ingress of large quantities of water into the gasification space.

Temperature

	A	B	C
Reaction temperature, °C	ca 1600	ca 1700	1100
Exit gas temperature, °C	100-120	not stated	200-400

VIII. Process development

A, B. Not stated

C. Laramie Energy Research Center, Wyoming, USA

Location

A. Angren (Uzbekistan, USSR)

B. Kemerovo (Kuznetsk, USSR)

C. Hanna (Wyoming, USA)

References

A. Angren

(1) M.M. Altschuler et al, Technical and Economic Analysis of the Angren, Yuzhno-Abinsk and Lisichank Underground Gasification Stations during 1963 Naucnye trudy VNII, Podzemgaza, No 13 (1965), pp 107-116 (UCRL-Translation 10807)

(2) M. Gubergrits et al, Manufacture of gas from combustible shales and by means of underground coal gasification, IGU/B 6, 1976, 13 World Gas Conference, London 1976

(3) W.W. Iwanow (Ivanov), Das Werk Angren für die Untertagevergasung von Braunkohle (The Angren installation for underground gasification of lignite), Podzemgaza Uglei 1957 No 3, pp 13-15 (Stbv-Übersetzung 328) (translation from Russian into German)

(4) N.V. Lavrov et al, Underground Gasification of the Angren Brown Coal Field, Khim. tverd. topl. No 1 (1971), pp 73-79 (UCRL-Translation 10922)

(5) N.V. Lavrov et al, Results of Studies of Underground Gasification of Brown Coal, Khim. tverd, topl. No 1 (1968), pp 113-120 (UCRL-Translation 10805)

B. Yuzhno-Abinsk

(1) See M.M. Altschuler, ref (1), under Angren

(2) O. de Combrugghe, La Gazeification Souterraine en URSS (Underground gasification in the USSR), Annal. d. Min. d. Belg., No 5 (1959), p 478

(3) M.J. Ustinov and P.A. Bukin, Der Einfluss der Hauptfaktoren auf den Vergasungsprozess im Baufeld Nr. 1 der Vergasungsanlage Jushno-Abinsk (Influence of the key factors on the underground gasification process in Field No 1 of the Yuzhno-Abinsk gasification installation), Podzemgaza Uglei, No 2 (1958), p 3 (Stbv-Übersetzung 561) (translation from Russian into German)

C. Hanna

(1) C.F. Brandenburg et al, In situ coal gasification - prospects as a source of utility fuel, August 1976, ERDA publication, Laramie Energy Research Center, Laramie, Wyoming

(2) D.D. Fischer et al, Status of the linked vertical well process in underground coal gasification, Paper FS4, 2nd Annual Underground Coal Gasification Symposium, 10-12 August 1976, Morgantown, W. Virginia

General

(1) P.V. Skafa, Podzemnaja gazifikazija ugley (Underground coal gasification), Gosgortechizdat, Moscow 1960 (UCRL-Trans 10880)

(2) J.L. Elder in Chemistry of coal utilization, H. Lowry (Ed), Supplementary Volume, J. Wiley & Sons, Inc, New York, London 1963

(3) J.H. Lamb, Underground coal gasification, Energy Techn. Rev. No 14, Noyes Data Corporation, Park Ridge New Jersey, 1977

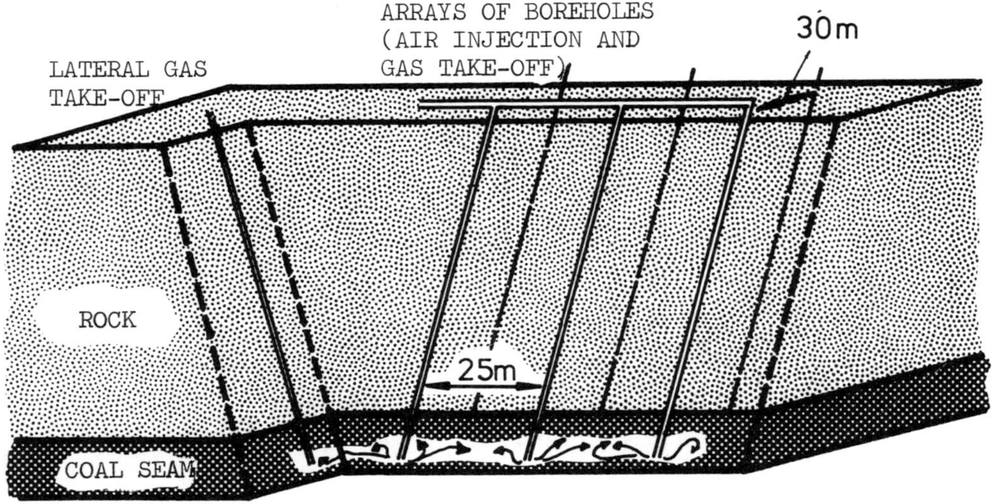

<u>Figure 42</u> Underground gasification at Angren: layout of the gasification field. The arrays of boreholes actually contain more borings than are shown. The possibility of lateral gas take-off is indicated.

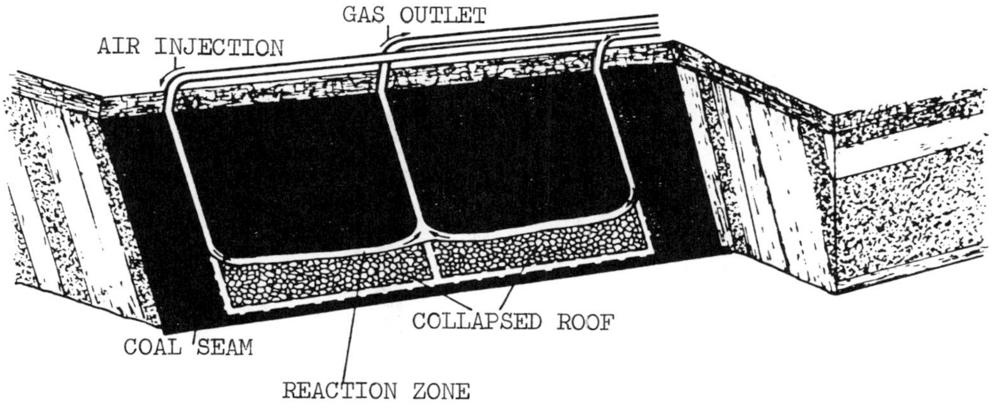

Figure 43   Underground gasification at Yuzhno-Abinsk: layout of the gasification field.

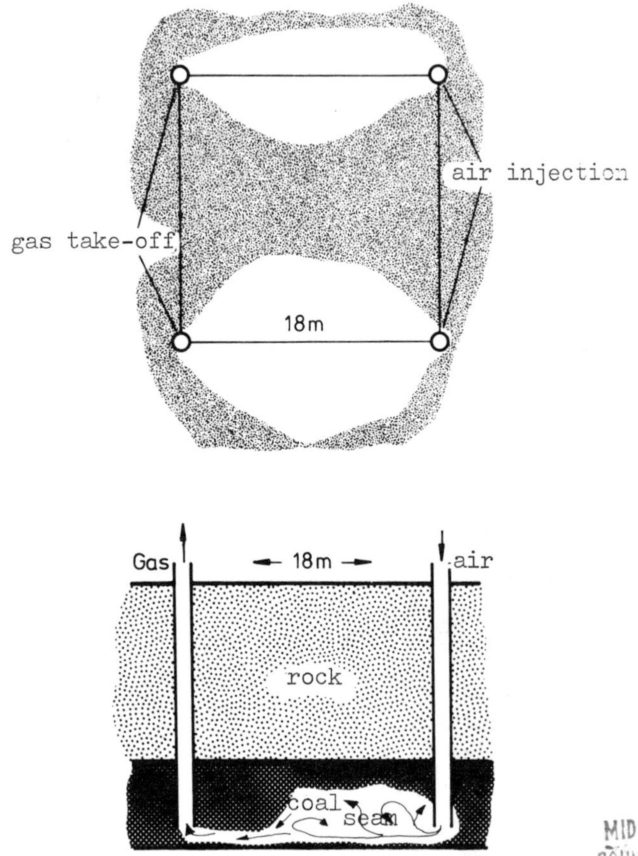

Figure 44 Borehole layout in the field experiment at Hanna, Phase II.
The upper diagram shows the ground plan of the gasification field; the lower one is a section through the field.